FUNCTIONAL MATHEMATICS

Level 1 and Level 2

Series editors: **William Rigby, Glyn Payne**

Authors: **Gwenllian Burns, Greg Byrd, Lynn Byrd, Harry Smith**

Part of Pearson

Longman is an imprint of Pearson Education Limited, a company incorporated in England and Wales, having its registered office at Edinburgh Gate, Harlow, Essex, CM20 2JE. Registered company number: 872828

www.pearsonschoolsandfecolleges.co.uk

Longman is a registered trademark of Pearson Education Limited

First published 2010
14 13 12 11 10
10 9 8 7 6 5 4 3 2

British Library Cataloguing in Publication Data
A catalogue record for this book is available from the British Library.
ISBN 978 1 408 26000 5

Edited by Elizabeth Bowden and Laurice Suess
Designed by Pearson Education Limited
Typeset by Tech-Set Ltd, Gateshead
Original illustrations © Pearson Education Ltd 2010
Illustrated by Tech-Set Ltd
Cover design by Wooden Ark
Picture research by Chrissie Martin
Cover photo/illustration © Getty Images
Printed in the UK by Scotprint

Acknowledgements

The author and publisher would like to thank the following individuals and organisations for permission to reproduce photographs.

Action Plus Sports Images: Steve Burdens 28-29; **Alamy Images:** Angela Hampton Picture Library 94, Blend Images 40-41, Brian Elliott 76-77, Duncan Hale-Sutton 104/2, ICP 32-33, Johnny Greig LSL 16-17, Justin Kase 78, Konstantinso Kokkinis 110, Mark Boulton 104/3, Paul Thompson Images 44b, Paul Underhill 72-73, Sam Toren 92, Wolfgang Polzer 92-93; **Construction Photography:** Ashley Cooper 104/1; **Corbis:** Kay Nietfeld / epa 30b, Lindsay Parnaby / epa 6c, Tim Parnell 10t; **Digital Vision:** 6t, 15; **Getty Images:** AFP / Joel Robine 88/5, National Geographic 108-109, Photographers Choice 102, 104-105, Stone+ 106, Workbook Stock 26; **Image Source Ltd:** 88/4; **Imagestate Media:** Phovoir 51, 115; **iStockphoto:** George Cairns 96-97, Susanna Fieramosca Naranjo 84-85; **Masterfile UK Ltd:** GardenPix 48-49; **Nature Picture Library:** Andy Rouse 36b; **Pearson Education Ltd:** Devon Olugbena Shaw 99, Gareth Boden 22, 82, 86, Ian Wedgewood 27, Jules Selmes 91, MindStudio 20-21, Rick Chapman 14; **Photolibrary.com:** Asia Images 80-81; **Photoshot Holdings Limited:** Thomas Schulze 88/6; **Plainpicture Ltd:** Briljans 42; **Press Association Images:** AP / David J Phillip 46; **Rex Features:** Alban Donohoe 36-37, John Powell 32c, SIPA Press 10c, Tony DiMaio 34; **Science Photo Library Ltd:** Ted Kinsman 44-45; **Shutterstock:** bikeriderlondon 38c, Brendan Howard 68, Brian A Jackson 38t, Cheryl Casey 67, Dmitry Pistrov 24-25, Fonats 103, Ibeth IA 88/1, Irina Rogova 8/4, Jeff Metzger 20, Karola I Marek 4, Kgelati 95, Kris Jacobs 18, Kristian Sekulic 83, Kzenon 100-101, Lana Langlois 36t, Lazar Mihal-Bogdan 88/2, MDD 70, Michael Stokes 88/3, Monkey Business Images 12-13, Olga & Elnar 39, Pedro Salverria 105, pzAxe 8/3, Robert Anthony 2, Rovenko Design 88-89, Sergey Peterman 8/2, Skyline 8/1, Sylvaine Thomas 17, Tatonka 74, Thomas Sztanek 98, Vuk Nenezic 30t, Wolfgang Amri 8/5, Yuri Arcurs 8-9, 76t, ZTS 8/6; **STILL Pictures The Whole Earth Photo Library:** Mark Edwards 108t; **SuperStock:** Age Fotostock 112-113; **TruEarth Imagery / Terrametrics:** 108b

Every effort has been made to contact copyright holders of material reproduced in this book. Any omissions will be rectified in subsequent printings if notice is given to the publishers.

All about Functional Maths

This book has been written to help you succeed in your **AQA Functional Maths** exams. The questions and practice papers in this book reflect the new **2010 Specification**, which has changed since the pilot phase. Throughout we provide every support in explaining functional questions, providing real-world contexts and exam-realistic practice.

Why learn Functional Maths? Functional Maths is about developing the skills you need to succeed in your job and in your life. In Functional Maths, you will practice your planning and problem solving skills and apply them to real-life contexts.

Examples of functional maths

- Planning a holiday budget
- Working out how long a train journey will take
- Working out a customer's invoice and adding VAT
- Arranging the furniture in your bedroom
- Working out how many songs you can fit on your MP3 player
- Understanding statistics in GCSE Geography

About the Levels

Level 1	Level 2
Functional Maths Level 1 is at the same difficulty level as Grade F at GCSE or National Curriculum Level 4. If you are taking a Foundation Diploma you will need to pass the Functional Maths exam at Level 1.	Functional Maths Level 2 is at the same difficulty level as Grade D/C at GCSE or National Curriculum Level 6. If you are taking a Higher or Advanced Diploma you will need to pass the Functional Maths exam at Level 2.

Remember…

If you are taking GCSE Maths, there will be **functional elements** in the exam questions. Studying for the Functional Maths qualification will help you prepare for these.

If you are on an Apprenticeship or Foundation Learning programme you will need to pass a Functional Maths exam. Your course-leader will tell you which level you need to take.

About the exam

There are separate Functional Maths exams for Level 1 and Level 2. The result of these exams will be delivered as a simple pass or fail. You do not need to have passed the Level 1 exam before taking the Level 2 exam.

Each AQA Functional Maths exam will last 1 hour and 30 minutes. You are allowed to use a calculator for the whole exam.

Your centre will be sent one or two of the **data sheets** four weeks before your exam. This information will be helpful for some of the questions in the exam. See page 52 or 116 for more details.

About Longman and how we can help you

This book provides you with all the mathematical skills and practice you need to succeed in your AQA Functional Maths exam at Levels 1 and 2. It is designed to be used as a stand-alone resource, or in combination with your GCSE Maths course. For the latter, it works perfectly in combination with Longman's **AQA GCSE Foundation Student Book**. See page viii for more details.

Our unique **Practise the maths** pages in each chapter will help you check and revise your basic maths skills. Our **Functional** pages provide real-life contexts and exam-realistic questions with which to develop and test those skills. Our innovative **Exam Café** provides focussed exam preparation. See pages vi and vii for more information about the features in this book.

Contents

Contents

LEVEL 2

How to use this book: Practise the maths

Every chapter starts with two pages of **Practise the maths** – a chance to check and revise the basic maths skills you will need for the functional questions in the chapter.

This book has been structured around mathematical topics, so you can build your skills and knowledge in sequence. Clear labelling of the topics makes them easy to look up for further practice, and easy to fit them into your GCSE course.

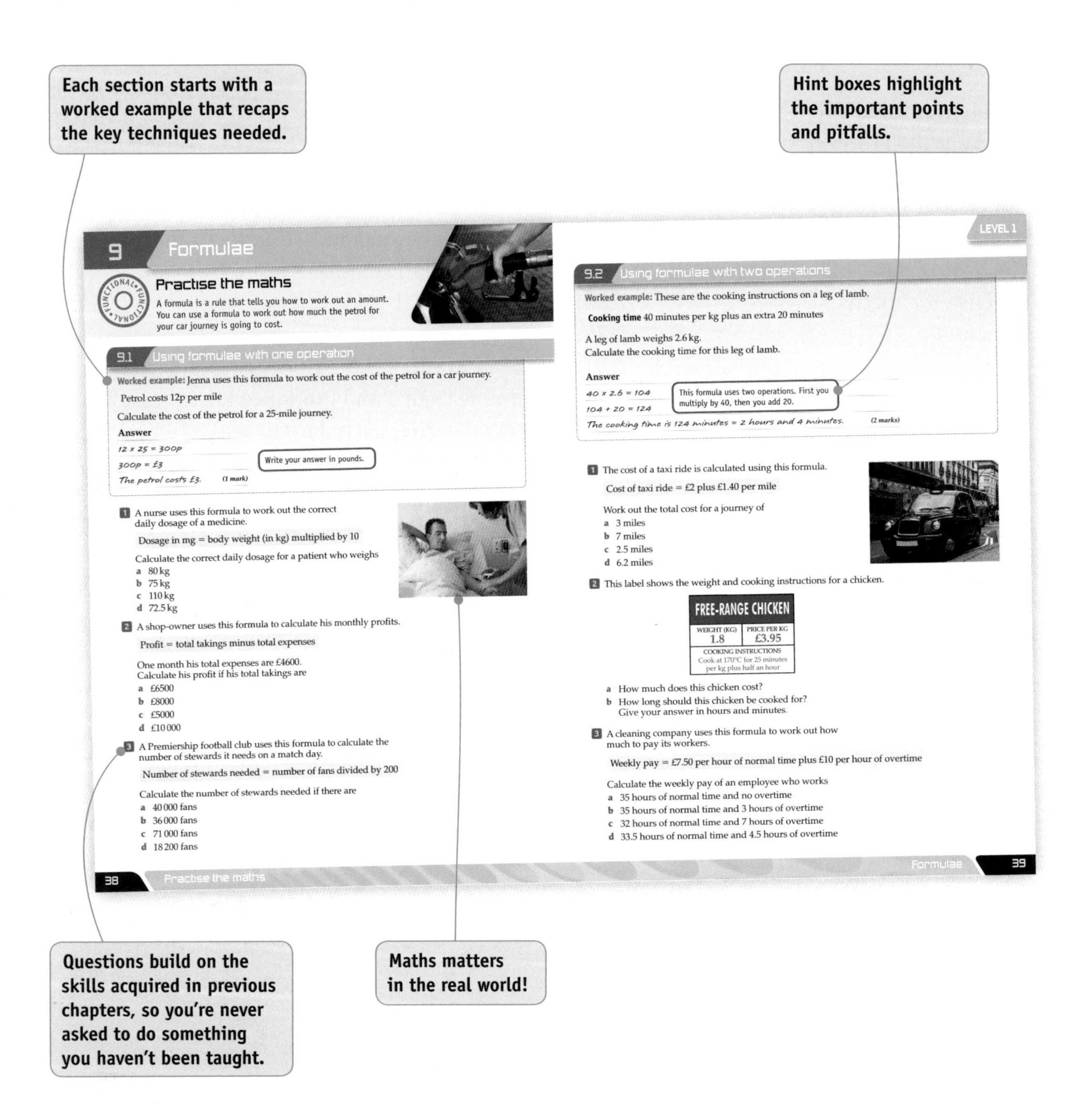

Each section starts with a worked example that recaps the key techniques needed.

Hint boxes highlight the important points and pitfalls.

Questions build on the skills acquired in previous chapters, so you're never asked to do something you haven't been taught.

Maths matters in the real world!

9 Formulae

Practise the maths

A formula is a rule that tells you how to work out an amount. You can use a formula to work out how much the petrol for your car journey is going to cost.

9.1 Using formulae with one operation

Worked example: Jenna uses this formula to work out the cost of the petrol for a car journey.

Petrol costs 12p per mile

Calculate the cost of the petrol for a 25-mile journey.

Answer

12 x 25 = 300p

300p = £3

The petrol costs £3. (1 mark)

Write your answer in pounds.

1 A nurse uses this formula to work out the correct daily dosage of a medicine.

Dosage in mg = body weight (in kg) multiplied by 10

Calculate the correct daily dosage for a patient who weighs

a 80 kg
b 75 kg
c 110 kg
d 72.5 kg

2 A shop-owner uses this formula to calculate his monthly profits.

Profit = total takings minus total expenses

One month his total expenses are £4600.
Calculate his profit if his total takings are

a £6500
b £8000
c £5000
d £10 000

3 A Premiership football club uses this formula to calculate the number of stewards it needs on a match day.

Number of stewards needed = number of fans divided by 200

Calculate the number of stewards needed if there are

a 40 000 fans
b 36 000 fans
c 71 000 fans
d 18 200 fans

38 Practise the maths

LEVEL 1

9.2 Using formulae with two operations

Worked example: These are the cooking instructions on a leg of lamb.

Cooking time 40 minutes per kg plus an extra 20 minutes

A leg of lamb weighs 2.6 kg.
Calculate the cooking time for this leg of lamb.

Answer

40 x 2.6 = 104

104 + 20 = 124

The cooking time is 124 minutes = 2 hours and 4 minutes. (2 marks)

This formula uses two operations. First you multiply by 40, then you add 20.

1 The cost of a taxi ride is calculated using this formula.

Cost of taxi ride = £2 plus £1.40 per mile

Work out the total cost for a journey of

a 3 miles
b 7 miles
c 2.5 miles
d 6.2 miles

2 This label shows the weight and cooking instructions for a chicken.

FREE-RANGE CHICKEN	
WEIGHT (KG) 1.8	PRICE PER KG £3.95
COOKING INSTRUCTIONS Cook at 170°C for 25 minutes per kg plus half an hour	

a How much does this chicken cost?
b How long should this chicken be cooked for? Give your answer in hours and minutes.

3 A cleaning company uses this formula to work out how much to pay its workers.

Weekly pay = £7.50 per hour of normal time plus £10 per hour of overtime

Calculate the weekly pay of an employee who works

a 35 hours of normal time and no overtime
b 35 hours of normal time and 3 hours of overtime
c 32 hours of normal time and 7 hours of overtime
d 33.5 hours of normal time and 4.5 hours of overtime

Formulae 39

How to use this book: Functional pages

Each Level starts with an introduction, providing guidance on **open questions** and the key skills in Functional Maths: **Representing, Analysing** and **Interpreting**.

See page 2 for Level 1 and page 66 for Level 2.

In each chapter, Functional pages provide real-life contexts and exam-realistic questions, which develop and test your mathematical skills. We have created a grading system to indicate the difficulty level of each question.

> **Beginner questions** This is where you should start.

>> **Improver questions**

>>> **Secure pass questions** These are the hardest questions, closely tailored to the exam style. When you're answering these correctly, you're on target!

Q Open questions are marked with this icon. Open questions have more than one possible correct answer. See page 5 and page 69 for more details.

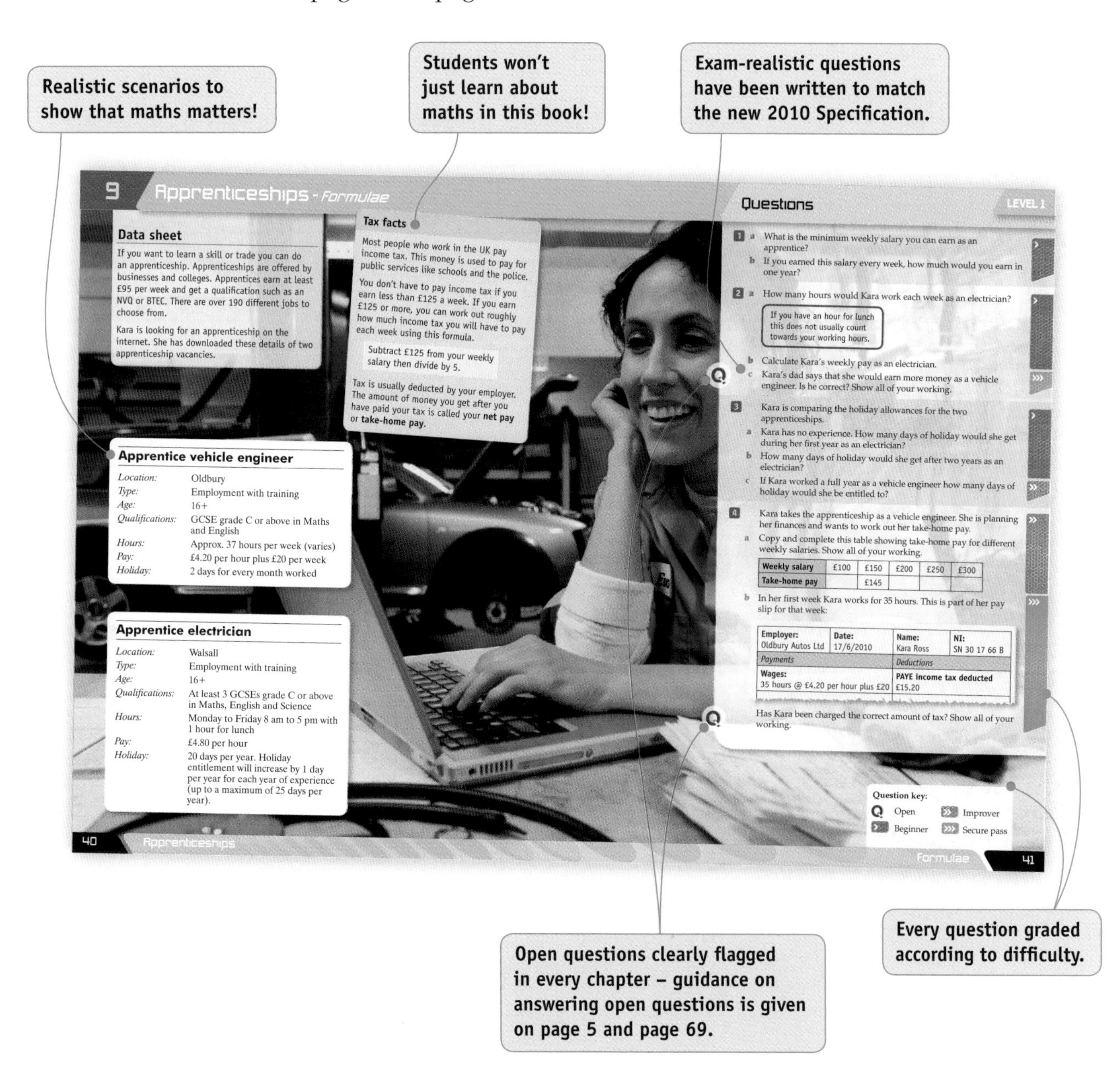

9 Apprenticeships - Formulae

Data sheet

If you want to learn a skill or trade you can do an apprenticeship. Apprenticeships are offered by businesses and colleges. Apprentices earn at least £95 per week and get a qualification such as an NVQ or BTEC. There are over 190 different jobs to choose from.

Kara is looking for an apprenticeship on the internet. She has downloaded these details of two apprenticeship vacancies.

Tax facts

Most people who work in the UK pay income tax. This money is used to pay for public services like schools and the police.

You don't have to pay income tax if you earn less than £125 a week. If you earn £125 or more, you can work out roughly how much income tax you will have to pay each week using this formula.

Subtract £125 from your weekly salary then divide by 5.

Tax is usually deducted by your employer. The amount of money you get after you have paid your tax is called your **net pay** or **take-home pay**.

Apprentice vehicle engineer

Location: Oldbury
Type: Employment with training
Age: 16+
Qualifications: GCSE grade C or above in Maths and English
Hours: Approx. 37 hours per week (varies)
Pay: £4.20 per hour plus £20 per week
Holiday: 2 days for every month worked

Apprentice electrician

Location: Walsall
Type: Employment with training
Age: 16+
Qualifications: At least 3 GCSEs grade C or above in Maths, English and Science
Hours: Monday to Friday 8 am to 5 pm with 1 hour for lunch
Pay: £4.80 per hour
Holiday: 20 days per year. Holiday entitlement will increase by 1 day per year for each year of experience (up to a maximum of 25 days per year).

40 Apprenticeships

Questions LEVEL 1

1 a What is the minimum weekly salary you can earn as an apprentice?
b If you earned this salary every week, how much would you earn in one year?

2 a How many hours would Kara work each week as an electrician?

If you have an hour for lunch this does not usually count towards your working hours.

b Calculate Kara's weekly pay as an electrician.
c Kara's dad says that she would earn more money as a vehicle engineer. Is he correct? Show all of your working.

3 Kara is comparing the holiday allowances for the two apprenticeships.
a Kara has no experience. How many days of holiday would she get during her first year as an electrician?
b How many days of holiday would she get after two years as an electrician?
c If Kara worked a full year as a vehicle engineer how many days of holiday would she be entitled to?

4 Kara takes the apprenticeship as a vehicle engineer. She is planning her finances and wants to work out her take-home pay.
a Copy and complete this table showing take-home pay for different weekly salaries. Show all of your working.

Weekly salary	£100	£150	£200	£250	£300
Take-home pay		£145			

b In her first week Kara works for 35 hours. This is part of her pay slip for that week:

Employer: Oldbury Autos Ltd	Date: 17/6/2010	Name: Kara Ross	NI: SN 30 17 66 B
Payments		*Deductions*	
Wages: 35 hours @ £4.20 per hour plus £20		PAYE income tax deducted £15.20	

Has Kara been charged the correct amount of tax? Show all of your working.

Question key:
Q Open
> Beginner
>> Improver
>>> Secure pass

Formulae 41

AQA GCSE Functional Maths Student Book Level 1	AQA GCSE Maths Foundation Student Book
Chapter/Section	Section
1 Working with whole numbers	
1.1	3.1, 3.3
1.2	11.5
1.3	11.1
1.4	11.2, 11.3
2 Recording data	
2.1	1.4
2.2	2.1
2.3	2.2
2.4	2.2
3 Fractions, decimals and percentages	
3.1	3.5
3.2	4.3
3.3	3.5, 4.1
3.4	15.2
4 Ratio	
4.1	4.5
4.2	4.5
5 Understanding data	
5.1	5.1
5.2	27.2
6 Mean and range	
6.1	6.1, 6.5
6.2	6.4, 9.1
7 Probability	
7.1	7.1
7.2	7.2, 7.3
8 Measures	
8.1	15.2, 15.4
8.2	23.2
8.3	23.3, 23.4
9 Formulae	
9.1	17.1
9.2	17.1
10 Perimeter, area and volume	
10.1	28.1
10.2	28.2
10.3	28.1, 28.2
10.4	28.3
11 Geometric shapes	
11.1	26.6
11.2	22.2
11.3	22.2, 24.2

AQA GCSE Functional Maths Student Book Level 2	AQA GCSE Maths Foundation Student Book
Chapter/Section	Section
12 Whole numbers and decimals	
12.1	3.3
12.2	11.5
12.3	12.1, 12.2, 12.3, 12.4
12.4	11.4
13 Recording data	
13.1	1.5
13.2	5.2
13.3	5.4, 5.5
14 Fractions, decimals and percentages	
14.1	3.3, 15.2, 15.4
14.2	3.5, 4.2, 4.3
14.3	4.1, 18.1, 18.2
14.4	14.3
15 Ratio	
15.1	4.5, 4.6
15.2	27.2
15.3	10.3
16 Understanding data	
16.1	5.3
16.2	5.4, 5.5
16.3	2.2, 2.3
17 Statistical methods	
17.1	6.2, 6.3, 6.4
17.2	6.2, 6.3, 6.4
17.3	9.4
18 Probability	
18.1	7.4, 7.5, 8.5
18.2	7.4, 7.5
19 Measures	
19.1	27.1
19.2	20.4
19.3	3.4
20 Formulae	
20.1	17.1
20.2	17.6
20.3	17.6
21 Perimeter, area and volume	
21.1	31.2
21.2	31.3
21.3	28.3, 31.4
22 Geometric shapes	
22.1	29.1
22.2	29.2

About the AQA Functional Mathematics series

As well as a Student Book, we are producing a **Teacher Guide** and an **ActiveTeach CD-ROM** in our AQA Functional Mathematics series:

FUNCTIONAL MATHEMATICS	 Student Book Levels 1 + 2 9781408260005	 ActiveTeach Levels 1 + 2 9781408259993	 Teacher Guide Levels 1 + 2 9781408260012

The **Teacher Guide** provides:

- **Lesson plans** for every double-page spread in the Student Book
- **Guided Practice Worksheets** for students who are struggling with the maths
- **Extension questions** for students who need stretch-&-challenge
- **Model answers** for the functional questions
- A further set of **Practice exam papers**, with guidance on the advanced release of data sheets

The **ActiveTeach CD-ROM** provides:

- The entire **Student Book on screen**, supported by the full range of ActiveTeach functionality
- Functional maths **video clips** – perfect as lesson-starters
- **Multi-lingual glossary**, giving written and audio definitions in English and five other languages: Bengali, Gujarati, Punjabi, Turkish and Urdu
- Functional problem-solving and multiple choice in our popular, interactive **Grade Studio**
- **Examiner Live** audio files, giving additional support to the functional questions
- Revision guidance and support in our innovative **Exam Café**

About the AQA GCSE Mathematics series – Foundation

This Student Book has been designed to work as a stand-alone resource or in combination with your GCSE course. For further learning and practice, you can refer to any of the Foundation level resources in our **AQA GCSE Mathematics 2010 series**:

STUDENT BOOK	PRACTICE BOOK	SUPPORT PRACTICE BOOK	TEACHER GUIDE with EDITABLE CD-ROM
Grades G-C 9781408232750	Grades G-C 9781408232736	Grades G-F 9781408240908	Grades G-C 9781408232743
ACTIVETEACH DVD-ROM	**PRACTICE BOOK - Digital Edition**	**SUPPORT PRACTICE BOOK - Digital Edition**	**ASSESSMENT PACK with EDITABLE CD-ROM - Covering all sets**
Grades G-C 9781408232729	Grades G-C 9781408243817	Grades G-F 9781408243824	Grades G-A* 9781408232842

Functional Maths is all about using maths in real-life situations. It involves taking mathematical skills you have already learned and applying them in everyday contexts, as well as in new and challenging situations.

This means Functional Maths is about more than just numbers – **it is about numbers and words**. To answer functional questions successfully, you will often need to provide written answers, conclusions and explanations. There's nothing to worry about – these are skills you use all the time in your English lessons. The purpose of this book is to show you how to use these skills in maths as well.

In order to use maths effectively, you must be able to understand information given to you, use it and interpret the results of your calculations. These are the functional maths skills that you need to develop and are called the **Process skills**.

The Process skills are divided into three sections: **Representing**, **Analysing** and **Interpreting**.

Process skills - Representing, Analysing and Interpreting

What is Representing?	What is Analysing?	What is Interpreting?
Representing means that you need to look at a problem, and work out what you need to do to solve it. You need to decide • what pieces of the information you need • what skills you need to use to solve the problem, e.g. addition, subtraction, multiplication etc • in what order you need to work things out.	Analysing means that you need to be able to use the skills you have decided you need to use, and work out the answers. You need to • work things out in the order you have already decided • use the mathematical skills already decided on to work out answers • check your calculations at every stage to make sure the answers make sense • write down all the calculations that you do and explain why you are doing them.	Interpreting means that you need to be able to use the answers to the calculations you have done. You need to explain what the answers mean and how they relate to the problem you've been asked to solve. You need to • explain what it is that you have worked out • explain how your answers relate to the problem you've been asked to solve • draw conclusions based on the comparisons that you have made.

Top tips – use DICE

- Decide - what is y–our plan? Write down every step.
- Information - write down everything that you use from the data sheet.
- Calculations - write them all down.
- Explain - what do your answers mean?

Process skills – Sample question and answer

Take a look at the question and model answer below.
It shows where the three different **Process skills** are used.
Here is the data sheet that provides all the information you need to answer the question.

Data sheet:

Calorie chart

Calories per 100 g of food			
roast potato	157	baked beans	64
baked potato	80	runner beans	24
white bread	233	yoghurt	95
wholemeal bread	216	apple juice (100 ml)	42
butter	750	orange juice (100 ml)	34
eggs	147	water (100 ml)	0
cheddar cheese	400	apple	45
cottage cheese	96	orange	35

Lunch options

Baked potato meal	
baked potato	300 g
baked beans	200 g
cheddar cheese	75 g
orange juice	300 ml
an apple	100 g

Egg sandwich meal	
wholemeal bread	200 g
eggs	100 g
butter	30 g
apple juice	300 ml
yoghurt	50 g

Question:

Alice is training for a marathon. She must make sure that she eats enough food each day to give her the energy she needs.
She does this by counting calories.
She decides to consume about 1000 calories for a lunch meal.
Her two favourite lunch meals are a baked potato meal or an egg sandwich meal.
Is the total number of calories for the baked potato meal more or less than 1000?

Model answer showing the three different Process skills:

Representing:
1. work out the number of calories in each ingredient of the baked potato meal using × and ÷
2. work out the total number of calories for the complete meal using +
3. decide whether the meal is more or less than 1000 calories

Analysing:
1. baked potato $80 \times 3 = 240$ calories
   baked beans $64 \times 2 = 128$ calories
   cheddar cheese $(400 \div 4) \times 3 = 300$ calories
   orange juice $34 \times 3 = 102$ calories
   1 apple 45 calories
2. total $= 240 + 128 + 300 + 102 + 45$
   $= 815$ calories

Interpreting:
3. 815 calories is less than 1000 calories

Why should I show my working?

In a Functional Maths exam it really matters if you **show your working**. Of course, it's best if you answer every question correctly, but that's not always possible. The point is, a wrong answer with no working will get no marks, while a wrong answer with lots of working can get nearly full marks.

Below is an example of part of a Level 1-style data sheet and question.

After the question you will see a model answer.

Then you will see three different solutions given by candidates.

All the candidates have the same wrong answer, but they each score a different number of marks. The examiner's comments show where the marks are given to each candidate.

Data sheet

The reasons for trips away from home within Scotland in 2007 and 2008

Purpose	2007	2008
Holiday, 1–3 nights	4.9 million	4.8 million
Holiday, 4+ nights	3.7 million	3.5 million
Visiting friends/relatives	1.9 million	1.6 million
Business	2.3 million	1.9 million
Total	**12.8 million**	**11.8 million**

Question:

What is the difference in the total number of holiday trips made within Scotland in 2007 and 2008? **(3 marks)**

Model answer:

Total number of holiday trips within Scotland in 2007 = 4.9 million + 3.7 million
= 8.6 million

Total number of holiday trips within Scotland in 2008 = 4.8 million + 3.5 million
= 8.3 million

Difference in the number of trips = 8.6 million − 8.3 million = 0.3 million

Candidates' answers		Examiner's comments
0.5 million	0 marks	Candidate has shown no working, the answer is wrong, I award no marks.
4.9 + 3.7 = 8.8 million *difference = 0.5 million*	1 mark	Candidate gets the first mark for showing 4.9 + 3.7, even though the answer is wrong. Candidate hasn't shown where the 0.5 comes from, so I award no more marks.
4.9 + 3.7 = 8.8 million *4.8 + 3.5 = 8.3 million* *8.8 − 8.3 = 0.5 million* *difference = 0.5 million*	2 marks	Candidate gets the first mark for showing 4.9 + 3.7, even though the answer is wrong. Candidate gets the second mark for showing that they are subtracting the two totals. Candidate doesn't get the last mark as the answer is wrong.

How should I answer an open question?

An **open question** is one where there may not be a single correct answer. There may be many different correct answers. We've indicated open questions throughout the book with this icon

When you answer an open question, think **DICE**.

- **D**ecide - what is your plan? Write down every step.
- **I**nformation - write down everything that you use from the data sheet.
- **C**alculations - write them all down.
- **E**xplain - what do your answers mean?

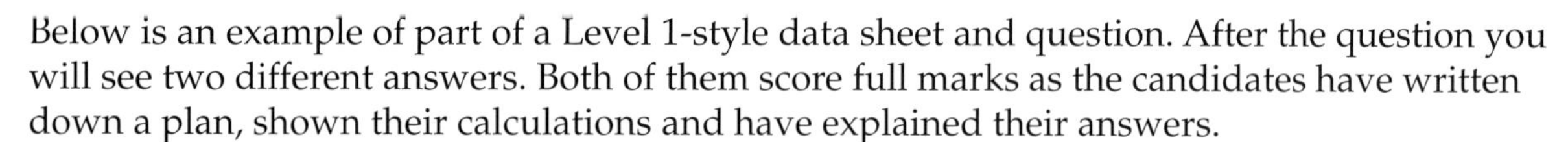

Below is an example of part of a Level 1-style data sheet and question. After the question you will see two different answers. Both of them score full marks as the candidates have written down a plan, shown their calculations and have explained their answers.

Data sheet

Edinburgh Tattoo ticket prices

Seating area	Price per person (£)
A	47
B	50
C, D, E	31
F	27
G	23

Edinburgh hotel prices

Hotel	Price per person per night (£)
Cameron Hotel	75
Castle Hotel	55
City Hotel	125

Question:

Moira is planning to stay in Edinburgh for two nights with her husband.

On one evening they want to go to the Edinburgh Tattoo.

They did the same last year, but took the cheapest options. This year they're hoping for better seats at the Tattoo, but cannot afford more than £360 for the tickets and the hotel.

Which hotel and Tattoo tickets do you recommend Moira books?

Candidates' answers:

Candidate A

Plan:
1. *work out the cost of the cheapest hotel*
2. *subtract the cost of the hotel from £360*
3. *divide the remainder of the money by 2 to work out the maximum amount that can be spent on one Tattoo ticket*
4. *explain my answer*

1. *Castle Hotel for 2 people for 2 nights = 55 x 2 x 2 = £220*
2. *£360 – £220 = £140 left to buy tickets*
3. *£140 ÷ 2 = £70*
4. *If Moira and her husband stay at the Castle Hotel they can afford any of the tickets for the Tattoo, as the most expensive ticket is £50.*

Candidate B

Plan:
1. *work out the cost of the medium price hotel*
2. *subtract the cost of the hotel from £360*
3. *work out if the money left over will buy Tattoo tickets*
4. *explain my answer*

1. *Cameron Hotel for 2 people for 2 nights = 75 x 2 x 2 = £300*
2. *£360 – £300 = £60 left to buy Tattoo tickets*
3. *tickets in area F = 2 x £27 = £54*
4. *Moira and her husband could stay at the Cameron Hotel and buy the second cheapest tickets for the Tattoo.*

1 Working with whole numbers

Practise the maths

To understand the information on climate change, you need to be able to use positive and negative numbers.

1.1 Using place value to round and order whole numbers

Worked example: Look at these numbers. 98 840 245 72 308 1255 783

a Write down the number that has eight hundreds.
b Write down the largest number. Give your answer in words.
c Round your answer to part **b** to the nearest hundred.
d Write the numbers in order, from smallest to largest.

Answer

a *840* (1 mark)
b *One thousand, two hundred and fifty-five* (1 mark)
c *1300* (1 mark)
d *72, 98, 245, 308, 783, 840, 1255* (2 marks)

> To round to the **nearest 100**, look at the digit in the tens column (12**5**5). If the digit is 5 or more, round up.

1 The average attendance at a home fixture for Liverpool Football Club is 43 508. Write the number 43 508 in words.

2 Round each of these numbers to the nearest 10.
a 34 **b** 123 **c** 359 **d** 1285

3 Write each set of numbers in order, from smallest to largest.
a 247, 427, 274, 386, 525 **b** 1350, 970, 1085, 1240, 1102

4 Write the correct sign, < or >, between each pair.
a 245 ___ 254 **b** 1345 ___ 1520 **c** £32 550 ___ £32 800

1.2 Working with negative numbers

Worked example: Arrange these temperatures in order, coldest first.
0 °C, −8 °C, 4 °C, −2 °C

Answer *−8 °C, −2 °C, 0 °C, 4 °C* (2 marks)

−10 −8 −6 −4 −2 0 2 4 6

> Use a number line to help you order positive and negative numbers. The further left you go, the lower the numbers are.

1 Write each set of temperatures in order, coldest first.
a 3 °C, −2 °C, −5 °C, 4 °C **b** −4 °C, 2 °C, 0 °C, −6 °C, −1 °C

2 This thermometer shows the temperature in Llandudno.

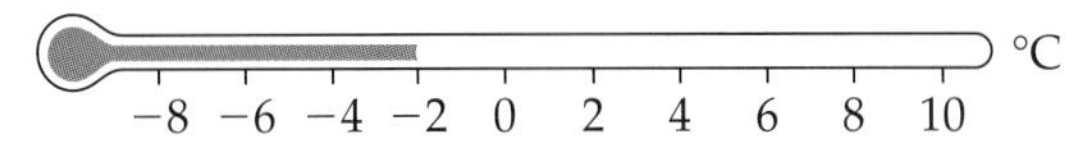

a What is the temperature in Llandudno?
b The temperature in Caernarfon is −4 °C. Which town is colder?

1.3 Adding and subtracting whole numbers

Worked example: Work out

a 128 + 145 **b** 462 − 138

Set out the calculations in columns. Always add or subtract the units column first.

Answer

a

```
  1 2 8
+ 1 4 5
  2 7 3
    1
```

b

```
  4 5̸6 1 2
- 1   3   8
  3   2   4
```

(1 mark)

2 is smaller than 8 so you need to exchange '10' from the tens column. The '60' in the tens column becomes '50'.

1 Kevin paid for a bus ticket with a £1 coin. The ticket cost 82p. How much change did he receive?

2 Work out

a 342 + 275 **b** 187 + 136 **c** 457 − 235 **d** 794 − 478

3 A shop had 94 tins of beans in stock one morning.
During the day they sold 58 tins and received a delivery of 144 tins.
How many tins of beans were there at the end of the day?

1.4 Multiplying and dividing whole numbers

Worked example: Work out

a 28×100 **b** $5 \div 10$ **c** 28×6 **d** $96 \div 6$

Answer

a 2800 *(1 mark)*

b 0.5 *(1 mark)*

To divide by 10, move the digits one place to the right.

c $20 \times 6 = 120$ *(1 mark)*

$8 \times 6 = 48$

$28 \times 6 = 168$

Partition 28 into 20 and 8, then multiply each by 6. Add the products to get the total.

d *(1 mark)*

```
   1 6
6)9 ³6
```

1 Work out

a 15×10 **b** 36×100 **c** 225×1000 **d** 108×100

2 Use a mental method to work out these.

a 18×4 **b** 26×8 **c** 43×7 **d** 75×6

3 Work out

a $460 \div 10$ **b** $3800 \div 100$ **c** $91 \div 7$ **d** $108 \div 6$

4 Tiles are sold in boxes of 100. Emily needs 1200 tiles.
How many boxes does she need to buy?

5 Dom needs £95 to buy a games console. He plans to save £5 each week. How many weeks will it take Dom to save enough money for the console?

Estimate → Calculate → Check it mate!

Data sheet

When choosing a mobile phone, you must decide between two payment options: Pay As You Go, or a monthly payment linked to a contract term (for example, an 18-month contract).

Whether Pay As You Go or a contract is better for you depends on how you use your phone.

Pay As You Go tariff	
Calls	20p per minute
Texts	10p per text
Minimum call charge	20p

Pay Monthly tariff	
Calls to landlines and Trideg mobiles	20p per minute
Calls to other network mobiles	35p per minute
Texts to other Trideg mobiles	3p per text
Texts to other network mobiles	12p per text
Minimum call charge	1 minute duration

18-month contract			
Online tariff	**Monthly cost**	**Free minutes**	**Free texts**
Trideg £10	£10	100	100
Trideg £15	£15	200	300
Trideg £18	£18	250	unlimited
Trideg £25	£25	500	unlimited
Note 1: Minutes and texts not used in a month cannot be carried over to the next month			
Note 2: Any usage over the set limits will be charged using the Pay Monthly tariffs			

Trideg Mobiles						
Model number	S246	S357	S12A	S408	S36B	S205
Recharge time (minutes)	190	120	150	180	210	160
Weight (g)	83	90	75	108	87	107
Cost	£10	£30	£45	£100	£80	£65

Questions

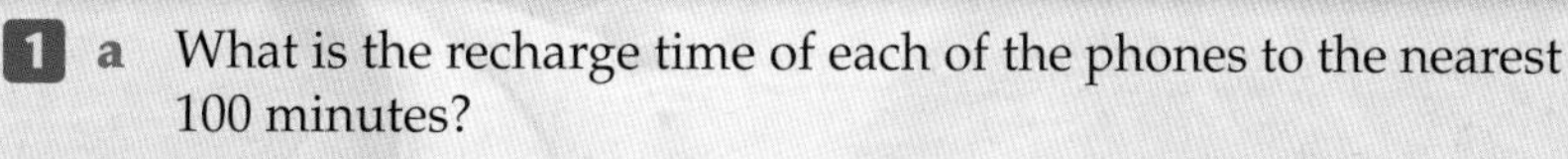

1 a What is the recharge time of each of the phones to the nearest 100 minutes?

b Which phone takes the least amount of time to recharge?

2 a Arrange the mobile phones in order of weight, starting with the lightest.

b Which is the heaviest phone?

c How much heavier is the S205 than the S12A?

3 Ella has a Pay As You Go package with her Trideg mobile.

a She paid £80 for her phone. Which model did she buy?

b She makes a call which lasts five minutes. Work out the cost of this call.

c How much will a call lasting 30 seconds cost?

d Ella has 125p of credit remaining. How many texts can she send?

e On Monday Ella had £20 credit on her phone. By Friday she has £9 of credit left. She has only used her phone to make calls. Estimate how many minutes of calls she has made. Explain why it is not possible to be sure how many minutes.

EXAM TIP

Give your answers in pounds and pence where appropriate.

4 Abdul uses his phone to keep in touch with his friends. All his friends have Trideg mobiles. Abdul signs an 18-month contract for the S357 phone, which is free as part of the deal. He chooses the £15 online tariff.

a During the first month, Abdul makes 130 minutes of calls and sends 200 texts. How much will he pay in total for month 1?

Check whether the phone usage is within the free limits of the contract.

b During the second month, he makes 240 minutes of calls and sends 200 texts. Abdul has exceeded his free call limit. How much will he pay for the extra minutes of calls?

5 Nora wants the S246 phone. She tells the Trideg adviser that she makes an average of 30 minutes of calls per month and sends 150 texts. None of Nora's friends have a Trideg mobile. The adviser tells her that she has two options.

Option 1: Pay As You Go

Option 2: 18-month contract, £10 online tariff, free phone

a Work out the cost of Option 1 over a 3-month period.

b Work out the cost of Option 2 over a 3-month period.

With Pay As You Go you must pay for the phone. However, the phone is free with an 18-month contract.

c Based on your answers in parts a and b, which option would you recommend to Nora?

6 Ibrahim uses his phone mainly for making calls. He makes an average of 280 minutes of calls per month. All of Ibrahim's contacts are part of the Trideg network.

Ibrahim is prepared to sign an 18-month contract to obtain the S408 for free. A Trideg adviser tells him that his best option is Trideg £25.

a Would you recommend the Trideg £25 package? Show all your working.

b Would your recommendation change if none of Ibrahim's contacts were part of the Trideg network? Show all your working.

Question key:

- Q Open
- › Beginner
- » Improver
- ››› Secure pass

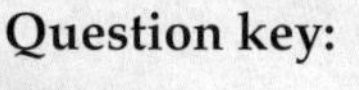

2 Recording data

Practise the maths

Radio stations collect data so that they can assess the age of their listeners. This helps them choose the type of music to play.

2.1 Recording data in a tally chart

Worked example: Elin asked a group of students how they travel to school. Here are their replies.

bike	car	walk	bus	bike	car	walk	car	bus	bus
bus	car	car	bike	bus	bus	bus	bike	bus	bus
walk	bike	car	car	bus	bus	bus	car	walk	walk

a Draw a tally chart to record this data.

b How many students' replies are recorded in your table?

Answer

a

Method of travel	Tally	Frequency
bike	𝍸	5
bus	𝍸 𝍸 II	12
car	𝍸 III	8
walk	𝍸	5

(3 marks)

A **tally chart** (or **frequency table**) has three columns: one for listing the items you are going to count, one for tally marks, and one to record the frequency of each item.

b $5 + 12 + 8 + 5 = 30$ students *(2 marks)*

1 Rose asked her friends to choose, from a list, which female singer they like best.
Here are their replies.

Shakira	Pixie Lott	Beyoncé	Pixie Lott	Beyoncé
Lily Allen	Shakira	Pixie Lott	Beyoncé	Beyoncé
Pixie Lott	Beyoncé	Beyoncé	Lily Allen	Lily Allen
Beyoncé	Lily Allen	Shakira	Pixie Lott	Beyoncé

a Draw a tally chart to record the data collected by Rose.

b Which singer was chosen the most often?

c How many of Rose's friends took part in the survey?

2.2 Representing data in a pictogram

Worked example: The table shows the numbers of music tracks downloaded by a group of friends.

Steven	Yossi	Albert
30	20	25

Draw a pictogram to represent this data. Use the symbol ● to represent 10 tracks.

Answer

Steven	●●●
Yossi	●●
Albert	●●◖

Key: ● = 10 tracks *(2 marks)*

The **key** shows how many items are represented by one picture. Always include a key with your pictogram.

1 Ibrahim collected this data on the numbers of text messages he sent.

Draw a pictogram to represent Ibrahim's data.

Use the symbol ✉ to represent 4 text messages.

Monday	Tuesday	Wednesday	Thursday	Friday
8	6	12	10	14

2.3 Representing data in a vertical bar chart

Worked example: Eloise collected this data on the favourite cricket grounds of a group of young cricketers.

Draw a vertical bar chart to represent Eloise's data.

Lord's	The Oval	Old Trafford	Trent Bridge	SWALEC Stadium
12	4	7	5	8

Answer

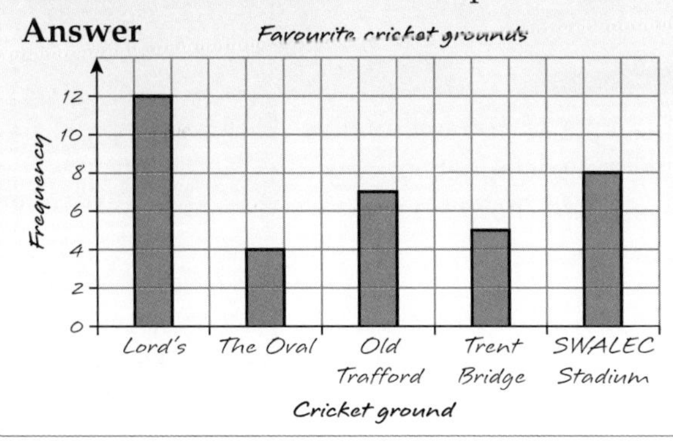

(3 marks)

The highest value is 12 so use a vertical scale from 0 to 14, going up in steps of 2.

Always label the axes and give your chart a title.

1 The table shows the ratings (out of 20) given to rides at Blackpool Pleasure Beach.

Draw a vertical bar chart to represent this data.

Grand National	Valhalla	Infusion	Avalanche	Bling
10	16	12	7	13

2.4 Representing data in a dual bar chart

Worked example: The table shows the numbers of e-mails and texts Heather sent over a period of five days.

Draw a dual bar chart to represent this data.

Day	Monday	Tuesday	Wednesday	Thursday	Friday
E-mails	4	2	8	6	5
Texts	7	5	3	10	13

Answer

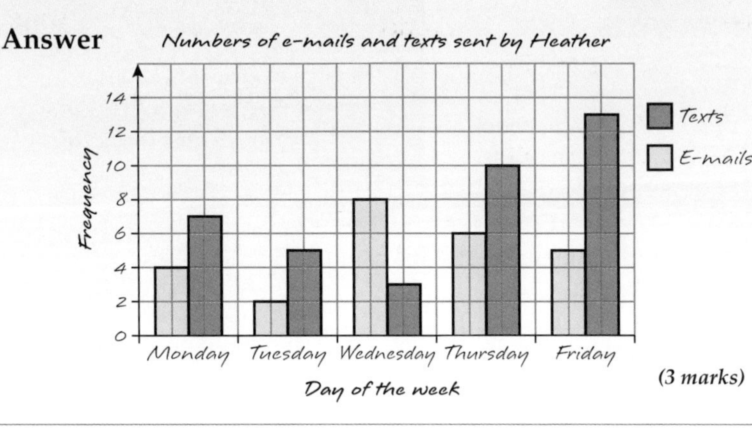

(3 marks)

In a **dual bar chart**, the bars are drawn side by side. Use a different colour for each item and include a key to explain what each colour represents.

1 The table shows the numbers of goals scored for and against some football teams.

Draw a dual bar chart to represent this data.

Team	For	Against
Liverpool	10	4
Stoke City	6	11
Arsenal	19	8
Aston Villa	6	4

Data sheet

Since their introduction, social networking sites such as Bebo, MySpace and Facebook have attracted millions of users. These sites offer people new ways of communicating via the internet, through either their computer or mobile phone.

These sites allow people to create their own online profile, detailing their interests, likes and dislikes.

Izzy and Arthur have conducted their own survey into social networking sites. Some of their results are given below.

Inbox: 1 new message!
Subject: Social networking trends

In August 2009 it was reported that the number of 15- to 24-year-olds who have a profile on a social networking site had dropped for the first time. In contrast, the number of 25- to 34-year-olds who regularly check up on sites had increased.

We asked a sample of 100 adults which social networking site they mainly use. The results are shown in the table below.

Social networking site	Number of adults
Bebo	15
MySpace	25
Facebook	50
other	10

We asked a sample of 100 adults how often they visit a social networking site. The results are shown in the table below.

Frequency of visits	Number of adults
every day	34
every other day	22
couple of times a week	18
once a week	22
less often	4

We asked a sample of 100 students, ages 8 to 17 years, which social networking site they mainly use.

Some of the results are shown below.

MySpace	Bebo	Bebo	Facebook	Bebo
Facebook	Facebook	Bebo	MySpace	Other
Other	Bebo	MySpace	Bebo	Other
MySpace	Bebo	Other	Bebo	Other
Bebo	Other	MySpace	Bebo	Facebook
MySpace	Bebo	Other	Other	Other

We asked a sample of 100 students and 100 adults their main reason for using social networking sites. The results are shown in the table below.

Reason	Number of students	Number of adults
stay in touch with friends I see regularly	34	22
look at my page or other people's pages or profiles	34	10
stay in touch with friends I rarely see in person	20	20
make new friends	10	8
other	2	40

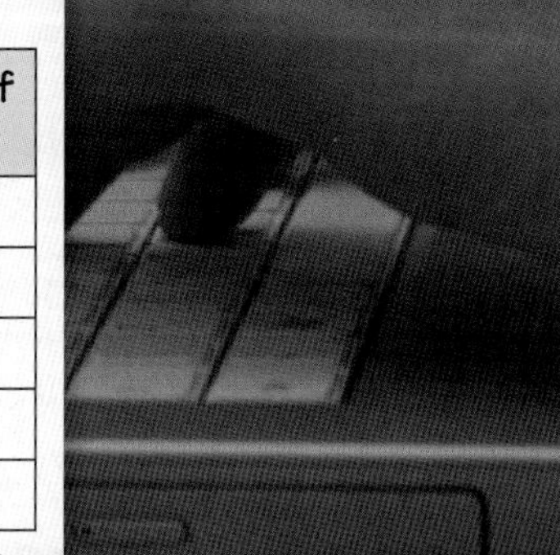

Questions

1 a Which social networking sites are identified in Izzy and Arthur's survey?

b 'Other' is listed as a social networking site. What does 'other' stand for?

2 a Draw a tally chart to record the data on the main networking sites used by students.

> Use three columns: 'Social networking site', 'Tally' and 'Frequency'.

b Which social networking site is recorded most often in your table in part **a**?

c How many students are included in your table?

d How many more students actually took part in this section of Izzy and Arthur's survey?

3 a Izzy and Arthur decide to use a pictogram to show the results from Q2**a**. Arthur wants to use • as the symbol to represent 4 students. Do you think this is a good symbol to use? Explain your answer.

b Draw a pictogram to show the results in your tally chart. Use the symbol ⊞ to represent 4 students.

4 a How many adults said Facebook was the main site that they visit?

b Draw a pictogram to show the social networking sites used by the adults. Use the symbol 🯅 to represent 5 adults.

5 Izzy and Arthur use their pictograms from Q3 and Q4 to compare the results for students and for adults. Is this a good idea? Explain your answer.

6 a Izzy draws a vertical bar chart to show the number of times the adults visit a social networking site. She uses a vertical scale from 0 to 40, going up in steps of 10. Do you think this is a sensible scale? Explain your answer.

b Izzy gives the bar chart the title 'Number of times people visit a social networking site'. Is this a good title? Explain your answer.

7 Draw a vertical bar chart to show the number of times the adults visit a social networking site.

8 When looking at people's reasons for using social networking sites, Izzy and Arthur have put all their results in one table. They also want to put the data for students and adults on a single chart. Which type of chart would you advise them to use? Explain your answer.

9 Draw a dual bar chart to show the reasons why both students and adults visit social networking sites.

10 Arthur says, 'The majority of students use social networking sites to communicate with other people'.

a Is Arthur correct? Explain your answer.

b Is Arthur's comment true for the adults. Explain your answer.

Question key:
- Q Open
- › Beginner
- ›› Improver
- ››› Secure pass

Practise the maths

Being able to convert between fractions, decimals and percentages means you can choose the best deal.

3.1 Working with fractions

Worked example:

a What fraction of this shape is shaded?

b Complete $\frac{7}{8} = \frac{\square}{16}$

Answer

a $\frac{7}{8}$ *(1 mark)*

Fraction $= \frac{\text{number of parts shaded}}{\text{total number of parts}}$

b $\frac{7}{8} = \frac{14}{16}$ *(1 mark)*

$\frac{7}{8}$ and $\frac{14}{16}$ are **equivalent fractions**, which means they have the same value. To find equivalent fractions, multiply the numerator and the denominator by the same number.

1 Write down the fraction of each shape that is shaded.

a

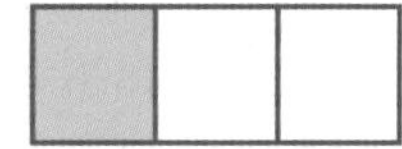

b

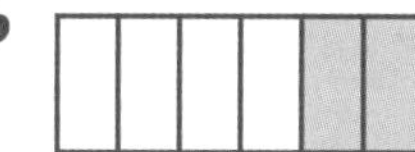

2 Copy and complete.

a $\frac{1}{2} = \frac{\square}{8}$ b $\frac{2}{3} = \frac{\square}{9}$ c $\frac{3}{4} = \frac{\square}{16}$ d $\frac{4}{5} = \frac{\square}{15}$

3.2 Converting between fractions, decimals and percentages

Worked example: Write

a 15% as a decimal b 27% as a fraction c $\frac{7}{10}$ as a decimal d 0.45 as a percentage e $\frac{1}{5}$ as a percentage.

Answer

a $15 \div 100 = 0.15$ *(1 mark)*

To convert a percentage to a decimal you divide by 100.

b $\frac{27}{100}$ *(1 mark)*

c $7 \div 10 = 0.7$ *(1 mark)*

d $0.45 \times 100\% = 45\%$ *(1 mark)*

e $\frac{1}{5} \times 100\% = 20\%$ *(1 mark)*

$\frac{1}{5} \times 100$ is the same as $100 \div 5$.

1 Copy and complete the table.

Fraction	Decimal	Percentage
$\frac{1}{2}$		
	0.25	
		75%
	0.2	
$\frac{7}{10}$		

3.3 Calculating a fraction and a percentage of an amount

Worked example: Work out

a $\frac{1}{2}$ of 120 **b** $\frac{1}{4}$ of 120 **c** 50% of 60 **d** 10% of 50 **e** 20% of 80

Answer

a $120 \div 2 = 60$ *(1 mark)*

b $120 \div 2 = 60$

$60 \div 2 = 30$

$(\text{Or } 120 \div 4 = 30)$ *(1 mark)*

c $60 \div 2 = 30$ *(1 mark)*

d $50 \div 10 = 5$ *(1 mark)*

e $80 \div 10 = 8$

$8 \times 2 = 16$ *(1 mark)*

50% = $\frac{1}{2}$, so to find 50% of an amount you divide by 2.

To find 10% of an amount, you divide by 10. You can use 10% to find other percentages. For example, 20% is two lots of 10%.

1 Work out

a $\frac{1}{2}$ of 40 **b** $\frac{1}{4}$ of 40 **c** $\frac{1}{2}$ of £280 **d** $\frac{1}{4}$ of £160

2 Work out

a 50% of 100 **b** 50% of 200 **c** 10% of £40 **d** 20% of 100

3 A package holiday costs £400. There is a 10% discount if the holiday is booked early.
How much is saved by booking the holiday early?

3.4 Adding and subtracting decimals

Worked example: Work out

a 12.8 + 5.6 **b** 24.6 − 13.25

Answer

a
```
   1 2 . 8
 +   5 . 6
 ---------
   1 8 . 4
     1
```
(1 mark)

b
```
   2 4 . 5̸6 1 0
 - 1 3 . 2  5
 ------------
   1 1 . 3  5
```
(1 mark)

Set out the calculations in columns and line up the decimal points.

Write an extra 0 in the hundreds column to help you line up the numbers correctly. This is called a **place holder.**

1 Work out

a 19.2 + 23.6 **b** 24.07 + 9.85 **c** 35.2 − 19.7 **d** 40.3 − 27.65

2 Lionel buys a CD for £8.49. He pays with a £20 note. How much change will he receive?

3 Amira cuts a 10 m piece of wood into three pieces. One piece is 2.54 m long. Another piece is 3.68 m long. How long is the third piece of wood?

Data sheet

The internet is becoming a very popular shopping location. In November 2009 alone, £5.3 billion was spent on online purchases in the United Kingdom.

However, in the same year, a consumer survey revealed that 50% of UK shoppers were fearful of shopping online because they were worried about credit card fraud.

Online shopping is particularly popular for items such as CDs, DVDs and computer games. Asif is searching on two computer game websites for games made by a company called Woo2 Games. The results of his searches are shown below.

Want to beat the rush for the January sales?

Having online stores means that retailers can start their sales early, even if their shops are closed for Christmas.
On Christmas Day 2008, a quarter of a million people logged on to Tesco Direct to shop for a bargain.

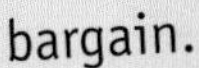

GAMES GALORE

Here are the results of your search for Woo2 Games:

Crazy Quad Bikes
£20.50 NEW
£16.00 USED

Football Funnies
£19.75 NEW

Brown Bear II
£16.50 NEW
£6.00 USED

Crazy Tractors
£26.99 NEW
£18.50 USED

Soldier's Return (JUST RELEASED)
£32.00 NEW

Planet Ziply
£14.50 NEW
£9.25 USED

Delivery rates per order		
Standard delivery	2–5 working days	£3.50
Express delivery	Within 24 hours	£4.95
Saver service	2–7 working days	FREE

Note: You must spend over £25 for the Saver service option.

THE NO.1 VIDEOGAME STORE

Here are the results of your search for Woo2 Games:

Crazy Quad Bikes
£16.50 NEW
£13.99 USED

Football Funnies
£14.75 NEW

Brown Bear II
£9.75 NEW
£6.25 USED

Crazy Tractors
£19.50 NEW
£16.00 USED

Soldier's Return (JUST RELEASED)
£32.00 NEW

Delivery rates per order		
Standard delivery	2–5 working days	£3.50
Express delivery	Within 24 hours	£4.95
Saver service	2–7 working days	FREE

Note: You must spend over £30 for the Saver service option.

Questions

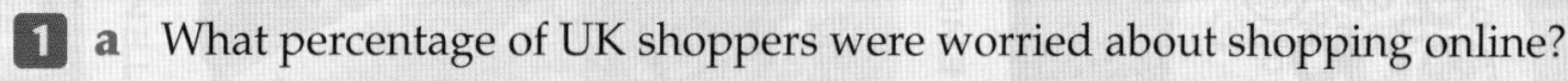

1 **a** What percentage of UK shoppers were worried about shopping online?
b Write your answer to part **a** as a fraction.

2 **a** How many people logged on to Tesco Direct on Christmas Day 2008? Write your answer in figures.
b What is a quarter as a fraction?
c Write your answer to part **b** as a decimal.

3 **a** Doris is shopping at Games Galore. She has a new copy of *Brown Bear II* and a used copy of *Crazy Tractors* in her online basket. What is the total cost of the contents of her basket?
b Does Doris qualify for Saver service delivery? Explain your answer.
c Doris wants her games within 5 days. Which delivery option would you recommend? Explain your answer.
d Work out the total cost of the order, using your delivery option from part **c**.

4 Jake has been given £80 for his birthday. He decides to buy *Soldier's Return* on the Games Galore site. He then uses the rest of his money to buy as many of the other games as he can. His budget must also cover any delivery charges.
Which other games could Jake buy?

5 Rupert has three games in his basket on The No.1 VideoGame Store site. One of the games is a new copy of *Crazy Quad Bikes*. The total cost of his basket contents is £42.25, not including the delivery charge.
What are the other two games in Rupert's basket?

Both Games Galore and The No.1 VideoGame Store have an end-of-season sale. The discounts are shown in the table below.

	Just-released titles	All other titles
Games Galore	Reduced by $\frac{1}{4}$	Reduced by 40%
The No.1 VideoGame Store	20% off	$\frac{1}{5}$ off

6 **a** What discount is The No.1 VideoGame Store offering on 'all other titles'? Give your answer as a percentage.
b What do you notice about The No.1 VideoGame Store discounts?

7 Elin wants to buy *Soldier's Return* for her son. Without working out the actual discount, advise Elin as to which site she should buy the game from. Explain your answer.

8 Dafydd says, 'The discounts offered by Games Galore are bigger than those offered by The No.1 VideoGame Store, so I will buy all my video games from Games Galore to save more money'.
Do you agree with Dafydd? Explain your answer.

Question key:
- Q Open
- > Beginner
- >> Improver
- >>> Secure pass

4 Ratio

Practise the maths

You need to use ratios in a lot of real-life situations – from mixing concrete to mixing ingredients for rock cakes!

4.1 Using ratio when one number is a multiple of the other

Worked example: A builder mixes sand and cement.
He uses 1 bag of cement for every 4 bags of sand.

a How many bags of sand does he mix with 3 bags of cement?

b How many bags of cement does he mix with 20 bags of sand?

First work out what one side of the ratio is multiplied by, then multiply the other side by the same number.

Answer

a *The ratio of cement : sand is 1 : 4.*

He uses 4 × 3 = 12 bags of sand. **(2 marks)**

b

1 : 4, ×5, ×5, □ : 20

He uses 1 × 5 = 5 bags of cement. **(2 marks)**

1 Alis makes an orange drink using the instructions on the bottle.

a How much water does she mix with 2 litres of orange squash?

b How much orange squash does she mix with 20 litres of water?

2 Beth is building a path using mortar and gravel.
She uses 3 bags of mortar for each bag of gravel.

a How many bags of mortar does she use with 2 bags of gravel?

b How many bags of gravel does she use with 12 bags of mortar?

First write down the ratio of mortar : gravel in numbers.

3 Ceri makes bracelets using beads.
She uses a pattern of 5 black beads for every 2 red beads.

a How many black beads does she use with 6 red beads?

b How many red beads does she use with 25 black beads?

First write down the ratio of black : red beads in numbers. Then work out what you must multiply 2 by to get 6.

4 Debbie is mixing orange paint.
She uses 5 litres of red paint for every 4 litres of yellow paint.

a How many litres of red paint does she use with 12 litres of yellow paint?

b How many litres of yellow paint does she use with 30 litres of red paint?

4.2 Solving problems by increasing or decreasing quantities at the same rate

Worked example: A recipe for rice pudding uses 100 g of rice.
The recipe is for 6 people.
How much rice is needed to make rice pudding for

a 12 people
b 3 people?

First work out which number you must multiply or divide the number of people by, then multiply or divide the ingredients by the same number.

Answer

a *6 people × 2 = 12 people*

100 g of rice × 2 = 200 g of rice **(2 marks)**

b *6 people ÷ 2 = 3 people*

100 g of rice ÷ 2 = 50 g of rice **(2 marks)**

1 A recipe for banana pudding uses 3 bananas.
The recipe is for 6 people.
How many bananas are needed to make banana pudding for

a 12 people
b 18 people
c 3 people
d 2 people?

2 The diagram shows part of a recipe for 15 doughnuts.

a How much margarine is needed for 30 doughnuts?
b How much flour is needed for
  i 3 doughnuts
  ii 6 doughnuts?

Makes 15 doughnuts
Ingredients:
200 g flour
50 g margarine

3 Sally is paid £36 for 4 hours' work.
How much is Sally paid for

a 8 hours' work
b 2 hours' work
c 5 hours' work
d 23 hours' work?

4 Sam makes decorations using beads and string.
For every 2 cm of string, Sam uses 2 blue beads and 1 yellow bead.

a How many blue beads does Sam use on a piece of string 10 cm long?
b How many yellow beads does Sam use on a piece of string 8 cm long?
c Sam uses 30 blue beads and 15 yellow beads on a piece of string.
How long is the piece of string?

In parts **a** and **c**, compare the length of string with the number of blue beads. In part **b**, compare the length of string with the number of yellow beads.

Data sheet

Hairdressers need to mix dyes for their customers every day. To do this they have to calculate the correct ratios of chemicals to use.

Louise is a hairdresser. When she colours a customer's hair, she must first find out what colour they already are and what colour they want to be.

She can then make a colouring mixture by mixing hair dye that matches the customer's hair colour with peroxide. She uses different peroxide concentrations for hair that is being made lighter or darker.

Peroxide percentage concentrations

Peroxide is a bleach that is mixed with water in different ratios to make the different percentage concentrations. The higher the percentage, the stronger the bleach.

The table below shows some commonly available peroxide solutions, which hairdressers can buy from suppliers. If they choose, hairdressers can mix these peroxide solutions with water in the correct ratio to make a weaker peroxide solution themselves.

Peroxide concentration	For hair that is being...
3%	made slightly darker
6%	kept the same shade or made 1 or 2 shades lighter
9%	made 3 shades lighter
12%	made 4 or 5 shades lighter

Hair shade chart

Shade 10	lightest blonde
Shade 9	very light blond
Shade 8	light blonde
Shade 7	medium blonde
Shade 6	dark blonde
Shade 5	light brown
Shade 4	medium brown
Shade 3	dark brown
Shade 2	very dark brown
Shade 1	black

Questions

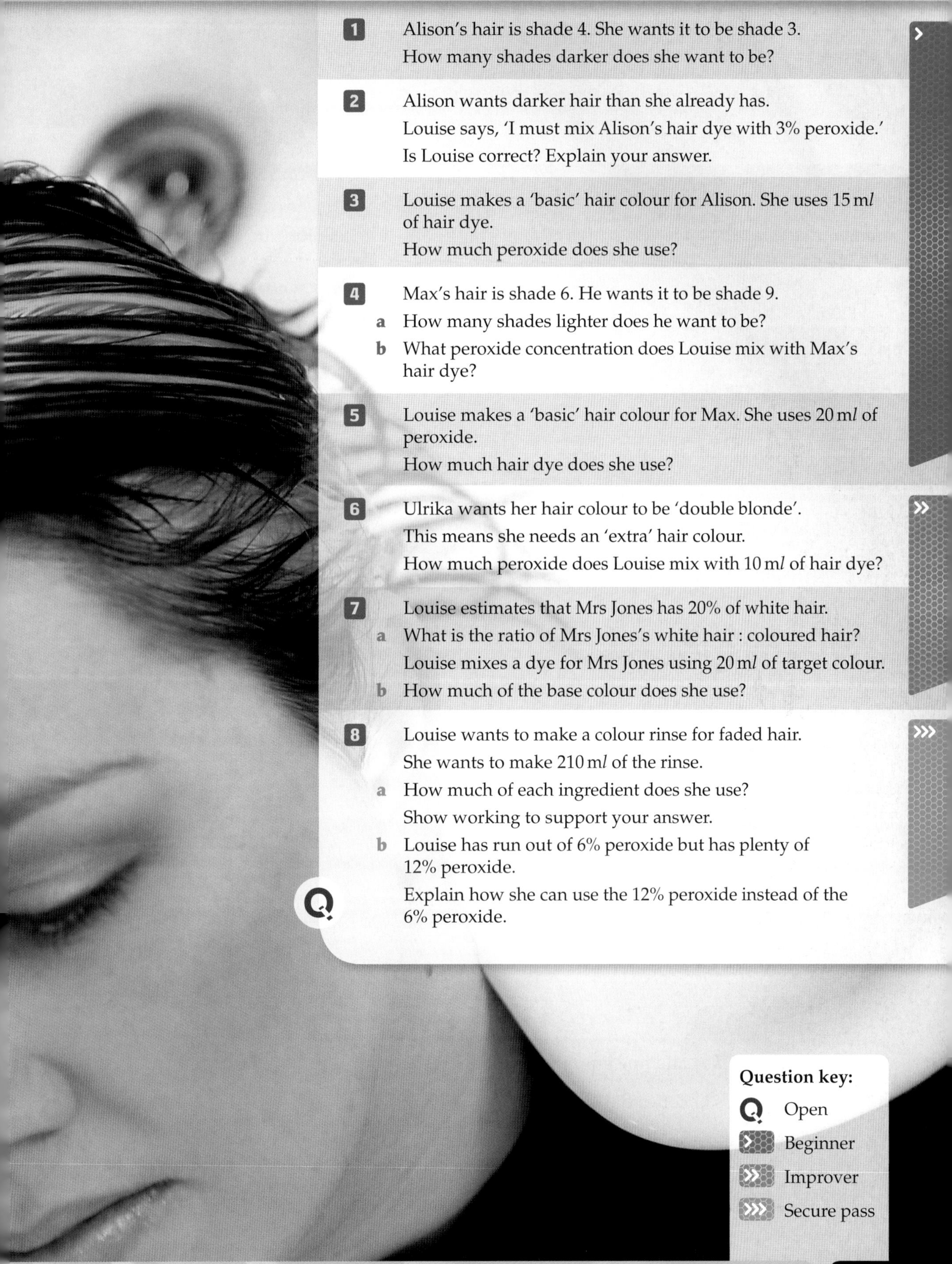

1 Alison's hair is shade 4. She wants it to be shade 3.
How many shades darker does she want to be?

2 Alison wants darker hair than she already has.
Louise says, 'I must mix Alison's hair dye with 3% peroxide.'
Is Louise correct? Explain your answer.

3 Louise makes a 'basic' hair colour for Alison. She uses 15 m*l* of hair dye.
How much peroxide does she use?

4 Max's hair is shade 6. He wants it to be shade 9.
a How many shades lighter does he want to be?
b What peroxide concentration does Louise mix with Max's hair dye?

5 Louise makes a 'basic' hair colour for Max. She uses 20 m*l* of peroxide.
How much hair dye does she use?

6 Ulrika wants her hair colour to be 'double blonde'.
This means she needs an 'extra' hair colour.
How much peroxide does Louise mix with 10 m*l* of hair dye?

7 Louise estimates that Mrs Jones has 20% of white hair.
a What is the ratio of Mrs Jones's white hair : coloured hair?
Louise mixes a dye for Mrs Jones using 20 m*l* of target colour.
b How much of the base colour does she use?

8 Louise wants to make a colour rinse for faded hair.
She wants to make 210 m*l* of the rinse.
a How much of each ingredient does she use?
Show working to support your answer.
b Louise has run out of 6% peroxide but has plenty of 12% peroxide.
Explain how she can use the 12% peroxide instead of the 6% peroxide.

Question key:
- Q Open
- > Beginner
- >> Improver
- >>> Secure pass

Practise the maths

Understanding data is important for working out which mobile phone tariff to choose or which car to buy.

5.1 Using and understanding data from tables, charts and graphs

Worked example: The table, pie chart and line graph show information about the sales at Coffee Café.

Coffee Café sales in 2008

Type of coffee	Number of cups (thousands)
cappuccino	525
latté	745
mocha	108
other	91

Coffee Café sales in 2009

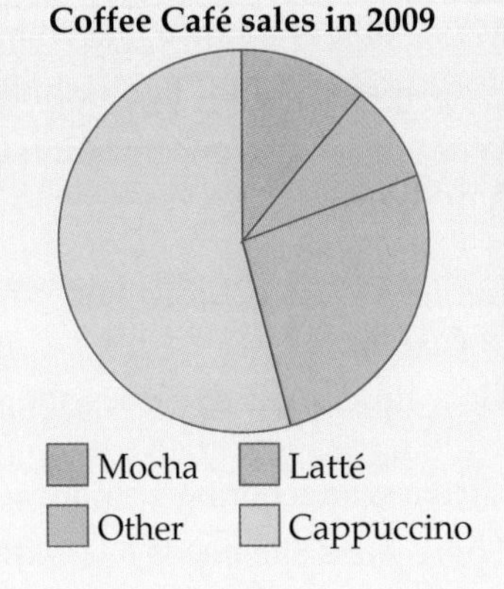

Coffee Café sales, 2004–2009

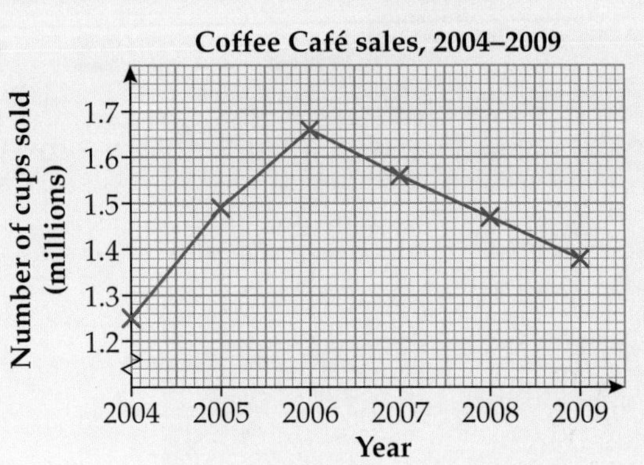

a How many cups of latté coffee did Coffee Café sell in 2008?

b Roughly what fraction of the coffees sold by Coffee Café in 2009 were latté?

c The total number of cups of coffee sold by Coffee Café in 2004 was 1.25 million. In 2009 the total number of cups sold was 1.38 million. This is an increase of 130 000 cups. Tim says, 'This means that Coffee Café's sales are improving.' Do you agree? Explain your answer.

Answer

a *Number of latté coffees sold in 2008 = 745 000* **(1 mark)** Use the table.

b *Roughly $\frac{1}{4}$ of the coffees sold in 2009 were latté.* **(1 mark)** Use the pie chart.

c *I disagree. The 2004 and 2009 numbers are correct, but the line graph shows that Coffee Café's sales improved from 2004 to 2006, but then they sold less and less each year.* **(2 marks)**

Use the table, pie chart and line graph from the worked example for Q1–3.

1 How many cups of coffee did Coffee Café sell in 2006?

2 What was the most popular type of coffee sold by Coffee Café in 2008?

3 **a** What major changes in the types of coffees sold do you notice from 2008 to 2009?
b Give a possible reason for your answer to part **a**.

4 This is Coffee Café's price list for 2010.

Abbie and her friends visit Coffee Café in 2010.

Abbie orders a regular latté with a vanilla shot.

Beth orders a regular espresso.

Chris orders a large mocha with cream and marshmallows.

Dylan orders a large mocha with cream.

Evan orders a regular Americano.

Fabio orders a regular espresso with an extra espresso shot.

What is the total bill?

Type of coffee	Regular	Large
Cappuccino	£2.45	£2.75
Latté	£2.05	£2.45
Mocha	£2.45	£2.75
Americano	£1.75	£1.95
Filter	£1.75	£1.95
Espresso	£1.45	£1.85
Macchiato	£1.45	£1.85
Shots (mint, vanilla or caramel) 40p		
Extra espresso 20p		
Cream 20p		
Marshmallows 20p		

5.2 Using scales on maps and drawings

Worked example: On a scale drawing of a building the scale used is '1 cm represents 2 m'.

a On the scale drawing the width of the building is 8 cm. How wide is the building in real life?

b In real life the length of the building is 30 m. How long is the building on the scale drawing?

Answer

a *1 cm represents 2 m, so the real width is*
8 × 2 = 16 m **(1 mark)**

b *1 cm represents 2 m, so the scale length is*
30 ÷ 2 = 15 cm **(1 mark)**

To go from the drawing to real life, multiply by the scale factor and change cm to m. To go from real life to the drawing, divide by the scale factor and change m to cm.

1 On a scale drawing of a building the scale used is '1 cm represents 5 m'.

a On the scale drawing the width of the building is 6 cm. How wide is the building in real life?

b In real life the length of the building is 45 m. How long is the building on the scale drawing?

2 The scale on this map is '1 cm represents 2 km'.

Kutlu walks from Norton railway station to the church at Little Gidding.

a Measure this distance on the map to the nearest mm.

b How far did Kutlu walk in real life?

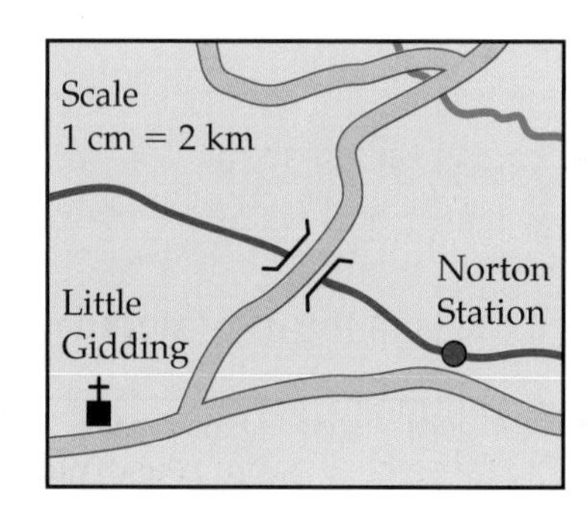

Data sheet

Gareth is a trainee fire prevention officer.
He writes a report on fires in the UK from 2001 to 2008.
He starts by looking at the causes of fires in the home in 2008.

Causes of building fires in the UK, 2001–2008

Year	Number of building fires (thousands)		
	Deliberate	Accidental	Total
2001	33	84	117
2002	30	82	112
2003	32	85	117
2004	32	81	113
2005	35	78	113
2006	31	74	105
2007	33	73	106
2008	28	70	98

Causes of fire in UK homes (2008)

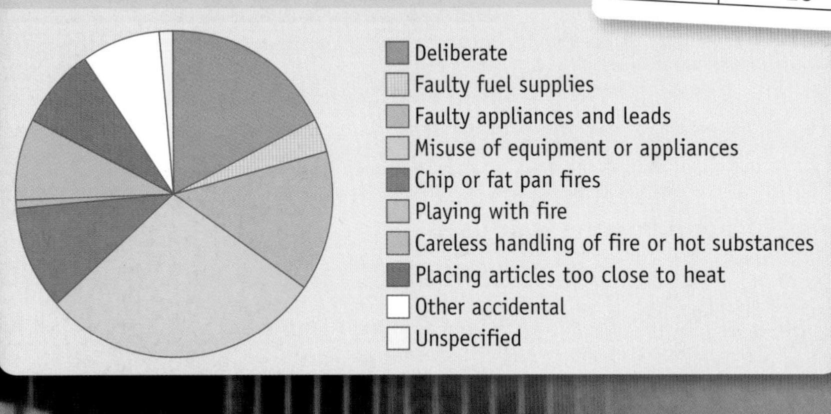

Total number of all fires in the UK, 2001–2008

Number of fires (thousands): 400, 450, 500, 550, 600

Year: 2001, 2002, 2003, 2004, 2005, 2006, 2007, 2008

Total number of false alarms in the UK, 2001–2008

Number of false alarms (thousands): 450, 500

Year: 2001, 2002, 2003, 2004, 2005, 2006, 2007, 2008

Questions

1 a Gareth says, 'Most of the fires in the home were caused by chip or fat pan accidents.'
Is Gareth correct? Explain your answer.

b Gareth also says, 'Faulty fuel supplies and faulty appliances and leads are more of a threat to UK homes than deliberate fires.' Is Gareth correct? Explain your answer.

2 Gareth looks at the numbers of accidental and deliberate building fires.

a How many of the building fires in 2003 were deliberate?

b Which year had the lowest total number of building fires?

c What is the difference between the numbers of accidental building fires in 2001 and 2008?

3 Gareth writes in his report, 'The number of deliberate fires was highest in 2005, with 35 000 fires. The number of deliberate fires was lowest in 2008, with 28 000 fires. This is a reduction of 7000 fires.'

Write a similar comment that compares the numbers of accidental fires in 2001 and 2008.

4 Gareth also includes in his report information on the total number of all fires and false alarms in the UK from 2001 to 2008.

a What was the total number of all fires in 2005?

b Which year had the highest total number of all fires?

c Between which two years did the number of false alarms not change?

5 Gareth writes in his report, 'One similarity between the sets of data is that overall the number of all fires and the number of false alarms have both gone down from 2001 to 2008.'

Is Gareth correct? Explain your answer.

6 Use the data to write down a prediction for the number of false alarms in 2009.

Explain your answer.

Question key:
- Open
- Beginner
- Improver
- Secure pass

6 Mean and range

Practise the maths

Designers of car and aeroplane seats use data from the public to work out the size of seats and the amount of legroom that is needed. They use the range and mean of people's heights, weights and waist measurements.

6.1 Working out the range

Worked example: Find the range for each of these sets of data.

a 4, 9, 8, 3, 5, 9, 10, 5, 5, 8

b

City	Cardiff	Glasgow	Sheffield	Newcastle	London
Minimum temperature (°C)	6	1	2	2	4

c

Shoe size	3	4	5	6	7	8	9	10	11
Frequency	12	15	17	24	32	24	20	10	6

Answer

a *Range = 10 − 3 = 7* *(2 marks)*

b *Range = 6 − 1 = 5°C* *(2 marks)*

c *Range = 11 − 3 = 8* *(2 marks)*

To work out the **range**, first find the highest and lowest values, then subtract the lowest from the highest.

1 Find the range for each of these sets of data.

a 12, 18, 9, 10, 8, 17, 14, 10

b 1.22, 1.81, 1.15, 1.77, 1.62, 1.88, 1.50

2 Work out the range for the number of mobile phones and the ages of lifeguards.

a

Number of mobile phones	Frequency
0	7
1	41
2	16
3	2

b

Age of lifeguard	Frequency
21	12
22	15
23	18
24	9

3 Two golfing friends compare their last eight golf scores.

John	72	89	77	70	69	85	83	87
Carlos	75	82	80	71	80	74	79	83

a Work out the range for John.

b Work out the range for Carlos.

c Who do you think is the more consistent golfer? Give a reason for your answer.

In golf, lower scores are better than higher ones, because the player has taken fewer shots to complete the course.

6.2 Working out the mean

Worked examples:

1 Find the mean for each of these sets of data.

a 1, 3, 5, 2, 4, 4, 1, 5, 2, 3

b

City	Cardiff	Glasgow	Sheffield	Newcastle	London
Maximum temperature (°C)	12	7	8	8	10

2 The total weight of five members of a basketball team is 355 kg.
What is the mean weight of the players?

Answers

1 a Mean $= \frac{1+3+5+2+4+4+1+5+2+3}{10} = \frac{30}{10} = 3$ **(2 marks)**

b Mean $= \frac{12+7+8+8+10}{5} = \frac{45}{5} = 9\,°C$ **(2 marks)**

2 Mean $= \frac{355}{5} = 71\text{ kg}$ **(2 marks)**

To work out the **mean**, first add together all the values, then divide the answer by the number of values that there are.

1 Find the mean for each of these sets of data.

a 17, 13, 15, 15, 10

b 3, 7, 6, 3, 8, 5, 4, 9, 7, 8

c 2, 1, 0, 2, 1, 3, 2, 1, 5, 0, 3, 4

2 This table shows the ages of five members of a football team.

Football team member	Alex	Bjorn	Carlos	Doyle	Ewan
Age (years)	18	21	17	22	17

Calculate the mean age of the five members.

3 Alan is 1.35 m tall. Brian is 0.18 m taller than Alan.
Chris is 6 cm shorter than Brian.
Work out their mean height.

4 Sandra has 8 chickens. In one week they produce a total of 48 eggs.
Work out the mean number of eggs produced by each chicken in one week.

5 The monthly wage bill for a factory where 20 people work is £46 000.
Work out the mean monthly wage.

6 Irena worked for 48 hours and earned £456.
How much does she earn per hour?

Data sheet

The London Marathon is a race over a distance of 26 miles and 385 yards around the streets of London. It has been held every year since 1981.

The Marathon is not restricted to elite runners. Thousands of ordinary men and women take part in order to raise money for charity, making it one of the UK's biggest fundraising events. In 2009, approximately 35 000 athletes took part, helping to raise a total of £47.2 million for charity.

There are three different starting points for the London Marathon. The starting points are coloured blue, green and red.

London Marathon start information

Start colour	Athlete numbers	Types of athlete
blue	151–28 000 54 251–59 000	elite, ballot, wheelchair
green	28 001–33 250	good for age, media
red	33 251–54 250	Gold Bond, overseas, guaranteed

Note: Charities can allocate Golden Bond places to athletes who pledge to raise money for them.

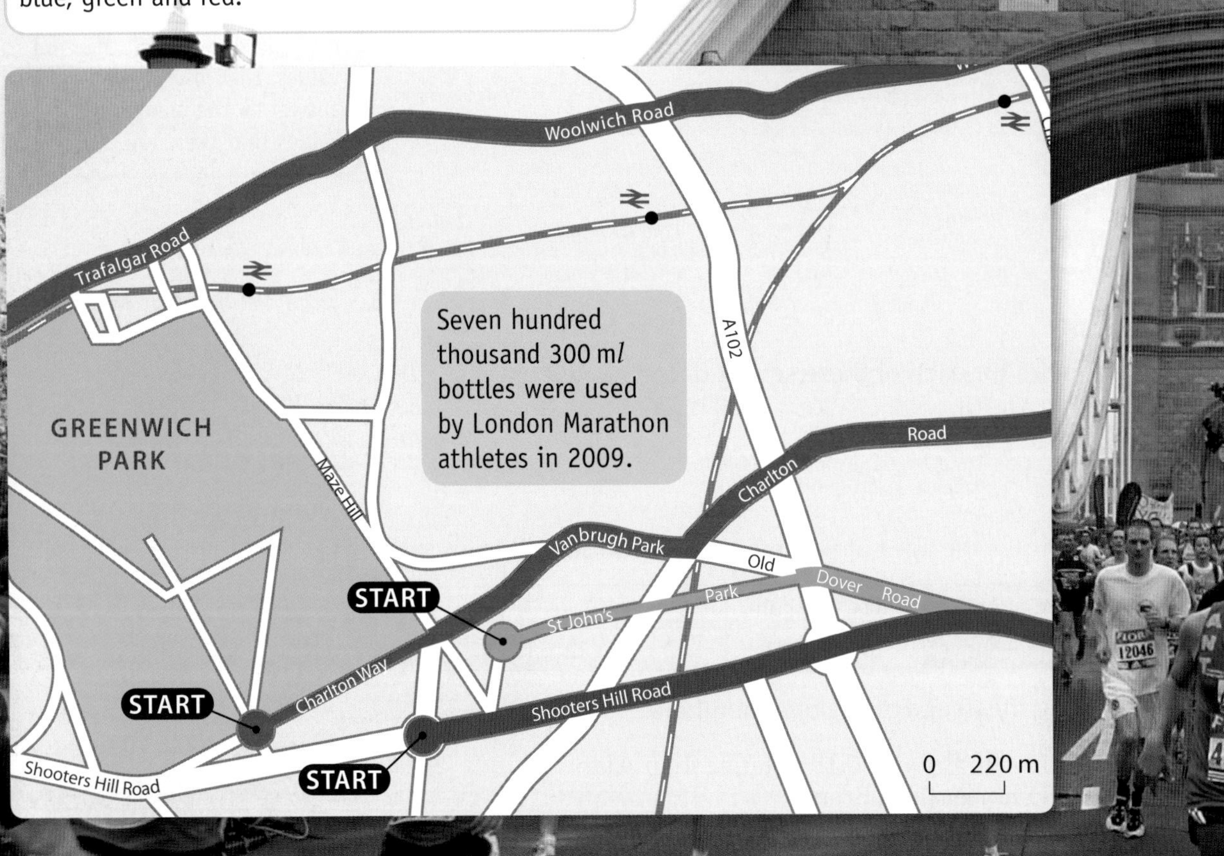

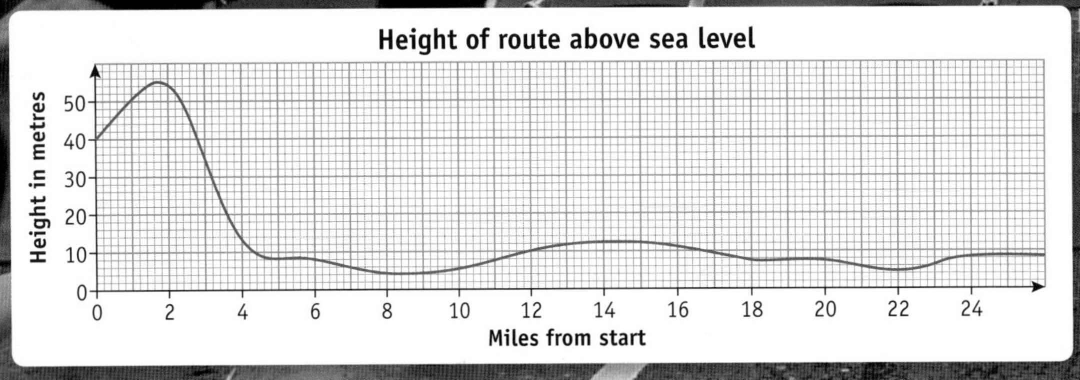

Questions

1 a Which park is the red start found by?
b Which road is the blue start found on?
c Which colour start does a wheelchair athlete use?

2 After how many metres do the athletes using the blue start join with the athletes using the green start?

3 a What is the height above sea level at mile 12 of the route?
b What is the range in height above sea level for the whole route?
c If you had to run two miles of the London Marathon, which two miles would you choose to run?
Explain your answer.

4 a What was the mean number of bottles of water used per athlete in 2009?
b What was the total amount of water used at the 2009 Marathon?
Give your answer in litres.

5 What was the mean amount of money raised per athlete in 2009?

6 In 2009, the average amount raised per athlete was £51 per mile.
Approximately 800 athletes started the race but didn't finish.
Estimate how much *more* money would have been raised if these athletes had finished the Marathon.
You must show all your working.

7 The table shows some information on the 2009 London marathon and the 2009 New York marathon. All times are given in hours (h), minutes (m) and seconds (s).

	London	New York
mean finishing time	4h 24m 31s	4h 24m 42s
fastest finishing time	2h 5m 10s	2h 9m 15s
slowest finishing time	10h 20m 47s	8h 59m 19s
number of men who finished the race	24 299	28 485
number of women who finished the race	11 351	15 175

Use this information to compare the two marathons.

Question key:
Q Open
> Beginner
>> Improver
>>> Secure pass

7 Probability

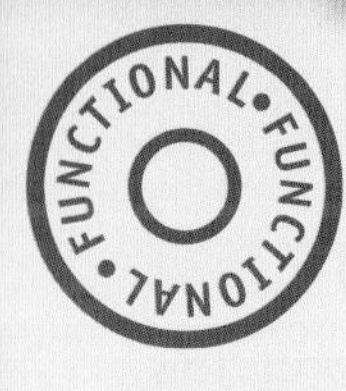

Practise the maths

If you can work out the probability of something happening, it could help you win money or – more importantly – stop you losing money.

7.1 Describing probability using words

Worked example: Choose a term from the box to describe the probability of each of these events.

impossible   very unlikely   unlikely   even chance   likely   very likely   certain

a Getting heads when you flip a fair coin.
b Getting a 2 when you roll a fair dice.
c Someone will run the 100 m race in less than 5 seconds at the next Olympic games.
d You will breathe out in the next 5 minutes.
e Picking the queen of hearts from a shuffled pack of cards.

Answer

a *even chance* **(1 mark)** — There are two possible outcomes (heads or tails). Both are equally likely.

b *very unlikely* **(1 mark)** — There are six possible outcomes (1, 2, 3, 4, 5 or 6). Each is equally likely.

c *impossible* **(1 mark)** — The fastest runners today can run 100 m in just under 10 seconds. No one can run twice as fast.

d *certain* **(1 mark)** — Unless you can hold your breath for 5 minutes!

e *very unlikely* **(1 mark)** — There are 52 cards in a pack (jokers are not included), so there are 52 possible outcomes. Each is equally likely. There is only one queen of hearts.

1 Choose a term from the box to describe the probability of each of these events.

impossible   very unlikely   unlikely   even chance   likely   very likely   certain

a Getting tails when you flip a fair coin.
b Getting a 6 when you roll a fair dice.
c Someone will break a world record at the next Olympic games.
d You will blink in the next 5 minutes.
e Picking the 2 of clubs from a shuffled pack of cards.
f Picking a club from a shuffled pack of cards.
g Picking a black card from a shuffled pack of cards.
h Someone picked at random is left-handed.
i You will live to be 120 years old.
j 1 April 2025 will be on a Monday.
k Getting an even number when you roll a fair dice.

2 Write down two events of your own that would have a probability of

a impossible
b very unlikely
c unlikely
d even chance
e likely
f very likely
g certain

7.2 Listing outcomes

Worked example: Graham has 2 white shirts, 4 blue shirts, 3 black shirts and 1 yellow shirt. Graham picks one shirt at random.

a List the possible outcomes for Graham.
b Which colour shirt is he most likely to choose?
c The probability that Graham chooses a black shirt is 0.3.
Show this probability on the probability scale.

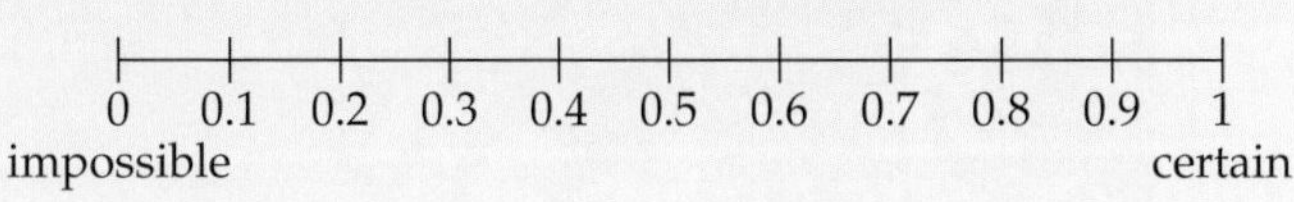

Answer

a *White, blue, black, yellow* *(2 marks)*

b *Blue* *(1 mark)*

Blue is the most common colour.

c

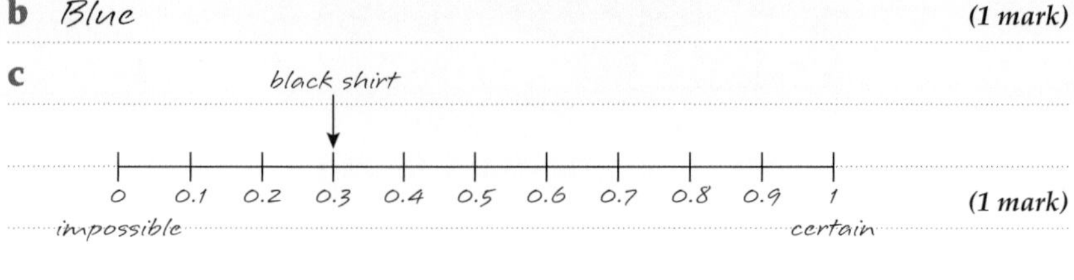

(1 mark)

1 List all the possible outcomes of each of these events.

a Flipping a coin.
b Rolling an ordinary dice.
c Picking a month of the year at random from a diary.
d Picking a day at random from a week.
e Rolling a four-sided dice.

2 Sam has 3 yellow t-shirts, 6 grey t-shirts, 1 blue t-shirt and 2 red t-shirts. She chooses one t-shirt at random.

a List all the possible outcomes for Sam.
b Which colour t-shirt is she most likely to choose?
c The probability that Sam chooses a grey t-shirt is 0.5.
The probability that Sam chooses a yellow t-shirt is 0.25.
Show these probabilities on a copy of the probability scale.

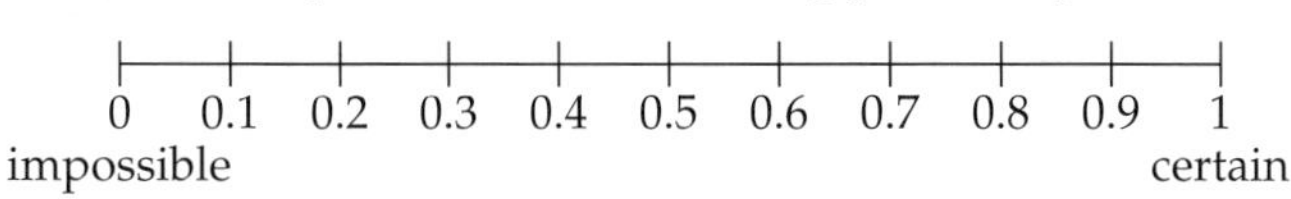

Data sheet

The National Lottery started in the UK in 1994. Since then, more than £24 billion has been raised for good causes, and more than £2 billion is being spent on the London 2012 Olympics.

People spend almost £50 million every week on buying lottery tickets.

How the money from National Lottery ticket sales is used

Where the money goes	Percentage of money from ticket sales
prize money	50%
government lottery duty	12%
shops that sell tickets	5%
operating costs	5%
good causes	28%

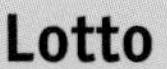

Lotto

There are lots of National Lottery games, but the main one is Lotto.

In the Lotto draw, players choose 6 numbers from 1 to 49 and a bonus ball number.

They win a prize if they have 3 or more matching numbers.

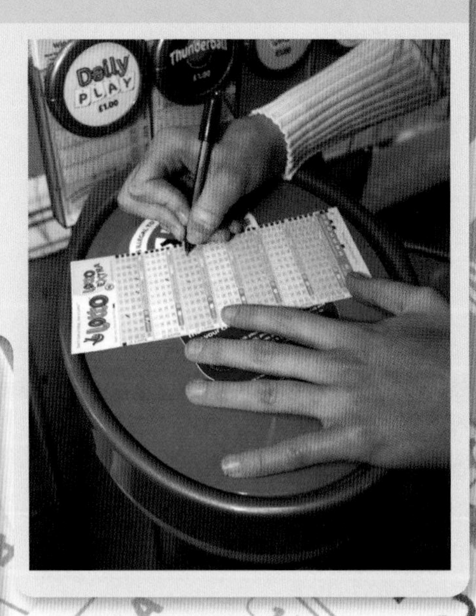

How the Lotto prize money is distributed

Players who match ...	Percentage of prize money
all 6 numbers	32%
5 numbers + bonus ball	10%
5 numbers	6%
4 numbers	13%
3 numbers	39%

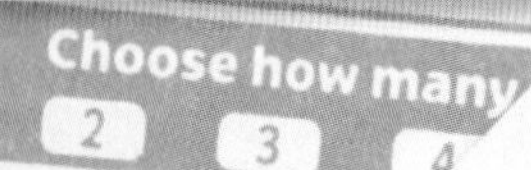

Probabilities of winning on the Lotto

How many matching numbers	Probability of winning	Typical prize money
6	0.000 000 07	£2 million
5 + bonus ball	0.000 000 4	£100 000
5	0.000 02	£1500
4	0.001	£65
3	0.02	£10

1 What fraction of the money from ticket sales is given as prize money?

2 What percentage of the money from ticket sales goes to the shops that sell the tickets?

3 What percentage of the Lotto prize money goes to people who have 4 numbers correct?

4 The table below shows how the £50 million in ticket sales each week is shared out.

Copy and complete the table.

Where the money goes	Percentage of money from ticket sales	Share of £50 million
prize money	50%	£25 million
government lottery duty		
shops that sell tickets		
operating costs		
good causes		

5 Copy and complete these sentences.

a The largest share of the Lotto prize money is given to those who have ☐ numbers correct.

b The smallest share of the Lotto prize money is given to those who have ☐ numbers correct.

6 Here is a probability scale. It shows that a probability of 1 means 'certain to happen' and a probability of 0 means 'impossible'.

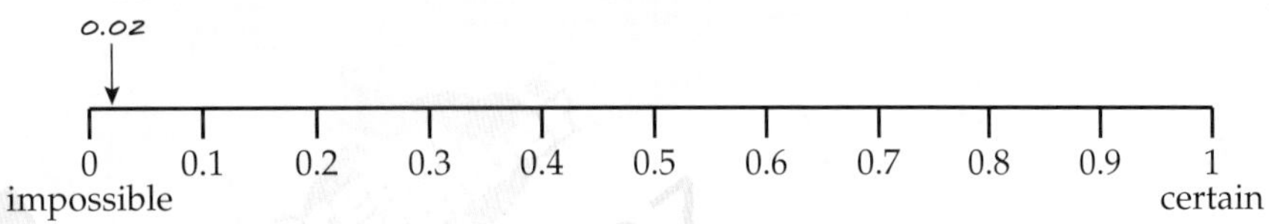

Arthur has drawn an arrow on the probability scale to show that the probability of winning £10 on the Lotto is 0.02.

Is it possible for Arthur to draw arrows on the scale to show the probabilities of winning the other amounts of money? Explain your answer.

7 Here are the probabilities of some events happening in the UK.

The probability that

A a pregnant woman will have twins is 0.03

B a road traffic accident is caused by a driver not looking properly is 0.35

C the cause of a plane crash is human error is 0.6

D a Scout chosen at random is a boy is 0.86

a Make a copy of the probability scale used in Q6.
Draw arrows on the scale to show the probabilities of the four events above.

b In the UK, which of these two events is more likely to happen?

A a pregnant woman will have twins

B a pregnant woman who buys a Lotto ticket will win £10

Explain your answer.

Question key:

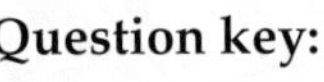

Q Open

› Beginner

›› Improver

››› Secure pass

8 Measures

Practise the maths

Money is needed in everyday life to buy goods and services. Understanding how to calculate with money helps you to choose the best-value options so you have more money left over. You also need to be able to measure lengths when doing DIY, or weights when cooking.

8.1 Calculating with money

Worked example: Work out

a £17.50 + £8.75 **b** £17.50 − £8.75 **c** £8.75 × 4 **d** £17.50 ÷ 5

Answer

a

```
  17.50
+  8.75
£26.25
  1 1
```

(1 mark)

b

```
  17.50
−  8.75
 £8.75
```

(1 mark)

c

```
   8.75
×     4
£35.00
  3  2
```

(1 mark)

d

```
  £3.50
5)17.²50
```

(1 mark)

1 Work out

a £12.95 + £3.50 **b** £105.99 + £26.75 **c** £72.50 − £18.25 **d** £30 − £12.75

e £6.95 × 3 **f** £145.95 × 2 **g** £9.60 ÷ 3 **h** £104.50 ÷ 5

2 Six friends go out for a meal.

a Anil orders a pasta dish costing £8.95, a pudding costing £3.95 and a drink costing £1.20. What is the total cost of Anil's meal?

b The total cost of the whole meal is £90.90. The friends decide to split the bill equally. How much do they each pay?

c How much more does Anil pay by splitting the total bill equally than by paying for his own meal?

8.2 Measuring and reading scales

Worked example:

a What is the length shown by this tape measure?

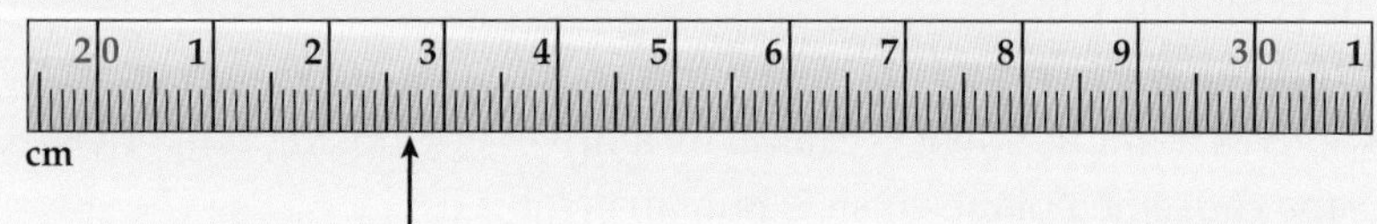

b Estimate the length of this blue line. ________________________

c Measure the length of the blue line above. Give your answer to the nearest mm.

Answer

a 22.7 cm (1 mark)

b 8 cm (1 mark)

c 8.2 cm (1 mark)

The width of a little finger is about 1 cm. The line is about 8 little finger widths long, so an estimate is 8 cm.

1 **a** Estimate the length of this blue line.

b Measure the length of the blue line. Give your answer to the nearest mm.

c Work out the difference between your estimate and the actual length of the blue line.

2 June is going on holiday to Spain by aeroplane.
She weighs her luggage at home.
The scale shows the weight of her luggage.

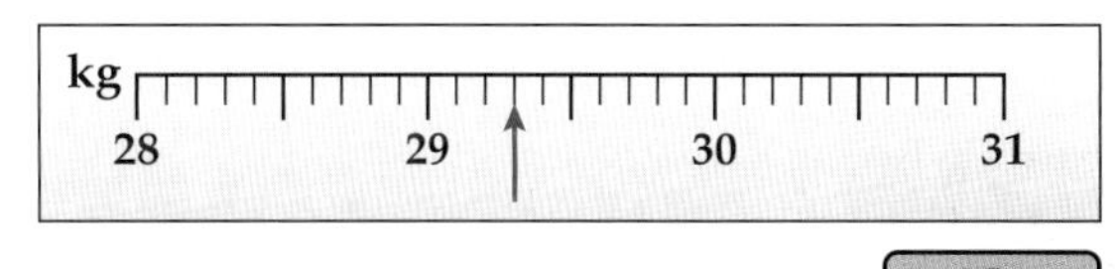

a What is the weight of her luggage?

b Passengers with luggage over 23 kg are charged £7.95 for every 1 kg or part of 1 kg it is over the 23 kg limit.
How much extra would June have to pay for her luggage?

> Find the difference between 23 kg and 30 kg, then multiply by £7.95.

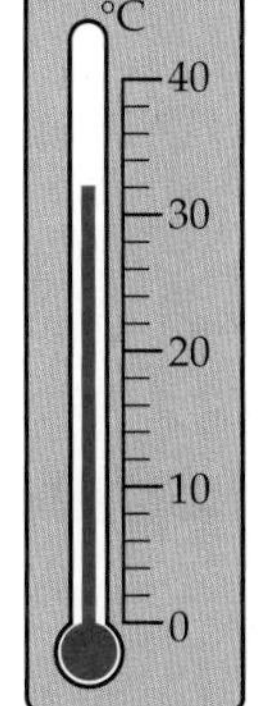

c June takes out some clothes and shoes, weighing a total of 3.8 kg.
How much does her luggage now weigh?

d How much extra would June now have to pay for her luggage?

e The thermometer shows the temperature when June arrives in Spain.
What is the temperature shown by the thermometer?

8.3 Working with time

Worked example:

a Write each of these times using the 24-hour clock.
i 6:30 am **ii** 6:30 pm

b My journey was meant to take 2 hours and 40 minutes but I was delayed for 35 minutes.
How long did my journey actually take?

Answer

> To write a pm time using the 24-hour clock you add 12 to the number of hours.

a **i** 06 30 *(1 mark)*

ii 18 30 *(1 mark)*

b 35 minutes = 20 minutes + 15 minutes

> Split the 35 minutes into 20 + 15. You need 20 minutes to add to the 40 minutes to make 1 hour.

2 hours 40 minutes + 20 minutes = 3 hours

3 hours + 15 minutes = 3 hours 15 minutes *(2 marks)*

1 Write each of these times using the 24-hour clock.
a 8:15 am **b** 9:30 pm

2 Write each of these times using the 12-hour clock.
a 04 25 **b** 13 20

3 My journey was meant to take 5 hours and 50 minutes but I was delayed for 45 minutes.
How long did my journey take?

4 My journey was meant to take 3 hours and 15 minutes but I arrived 25 minutes early. How long did my journey take?

> In Q4 you need to take the 25 minutes away from 3 hours 15 minutes.

5 Will must arrive at London Waterloo station by 3:30 pm.
The train journey from Ashford takes 40 minutes.
What is the latest train that Will can catch to arrive in time?

Data sheet

Every year in the UK more than five million wild animals and birds are injured in accidents. There are wildlife rescue centres throughout the UK which care for sick and injured animals, before releasing them back into the wild. One of the busiest centres treats 10 000 animals each year.

Amanda runs a wildlife rescue centre. When a hedgehog arrives at the centre she carries out a health check on it. She measures its waist circumference and its head-to-tail circumference.

A hedgehog is ready for release when

- its weight is 0.65 kg or more
- it has gained weight for at least 1 week
- its waist circumference is **more than** the value shown in the table, given its head-to-tail circumference.

Circumference	
Head-to-tail	**Waist**
35 cm	28.0 cm
36 cm	28.8 cm
37 cm	29.6 cm
38 cm	30.4 cm
39 cm	31.2 cm
40 cm	32.0 cm
41 cm	32.8 cm
42 cm	33.6 cm
43 cm	34.4 cm
44 cm	35.2 cm

Hedgehog number 78 – pre-release health check

Hedgehog number	78
Waist circumference	33.5 cm
Head-to-tail circumference	41 cm
Weight	675 g
Gained weight for at least one week?	✓

Average badger measurements

Weight in...	Adult male	Adult female
spring	8–9 kg	7–8 kg
autumn	11–12 kg	10–11 kg

Head-to-tail length ≈ 85–90 cm
Height to shoulder ≈ 30 cm

Questions

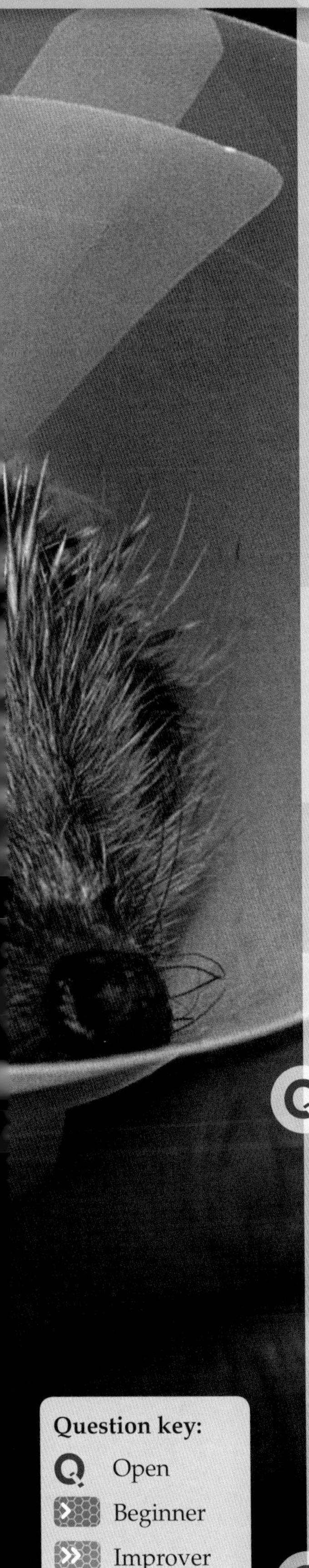

1 These tape measures show the waist circumference and the head-to-tail circumference of hedgehog number 78 on the day he arrived at Amanda's wildlife rescue centre.

Waist circumference

Head-to-tail circumference

Write down

a the waist circumference

b the head-to-tail circumference of hedgehog number 78.

Give your answers to the nearest mm.

2 Amanda also weighs the hedgehog.

The scales show the weight of hedgehog number 78 on the day he arrived at the centre.

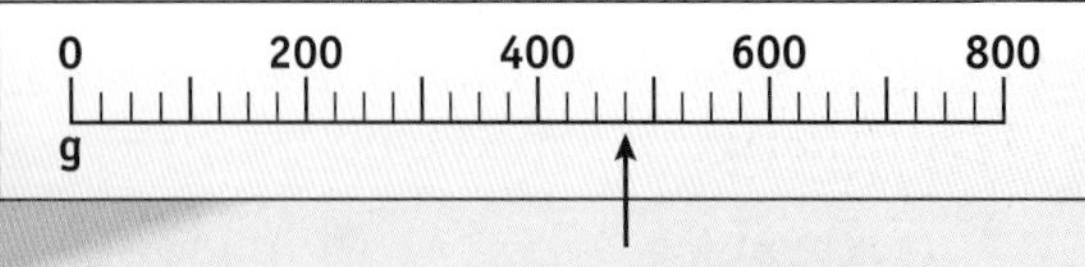

How much did he weigh on the day he arrived?

3 After two months at the centre, Amanda carries out a pre-release health check on hedgehog number 78.

Is he ready for release?

You must explain your answer and show all your working.

4 A hedgehog needs a cage with a minimum floor area of 0.36 m^2.

All the cages Amanda uses are either rectangular or square.

Write down the length and width of two possible cages that have exactly the minimum floor area.

Look at 'Practise the maths' 10.1 for how to calculate areas.

5 In the autumn, Amanda receives a phone call to say that two injured badgers are being brought into the rescue centre.

Amanda receives the phone call at 9:35 am.

The rescue team is 45 minutes away.

At what time should the badgers arrive at the centre?

6 The first badger is female and weighs 10.9 kg.

Is this badger within the average weight range?

7 The second badger is male. He weighs 6800 g.

He has a head-to-tail length of 525 mm.

Amanda says, 'This could be an adult male that is very underweight.'

a Is Amanda correct? Explain your answer.

b What other conclusion could Amanda make about this badger? Explain your answer.

Question key:

- Open
- Beginner
- Improver
- Secure pass

9 Formulae

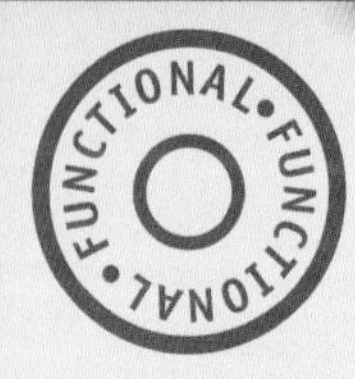

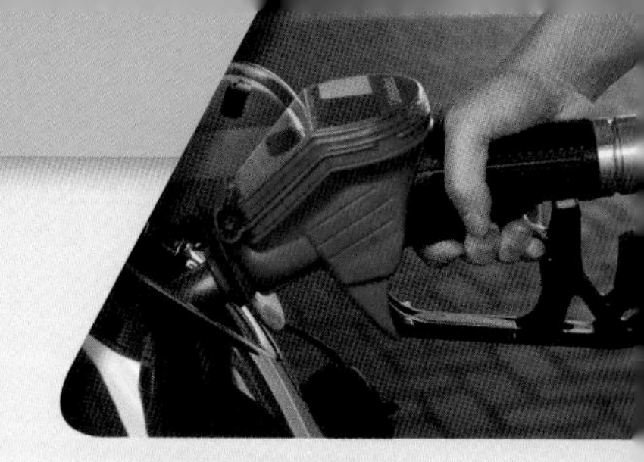

Practise the maths

A formula is a rule that tells you how to work out an amount. You can use a formula to work out how much the petrol for your car journey is going to cost.

9.1 Using formulae with one operation

Worked example: Jenna uses this formula to work out the cost of the petrol for a car journey.

Petrol costs 12p per mile

Calculate the cost of the petrol for a 25-mile journey.

Answer

12 × 25 = 300p

300p = £3

The petrol costs £3. *(1 mark)*

Write your answer in pounds.

1 A nurse uses this formula to work out the correct daily dosage of a medicine.

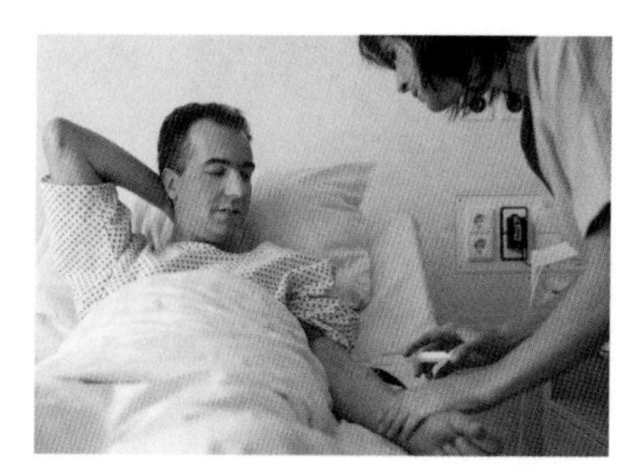

Dosage in mg = body weight (in kg) multiplied by 10

Calculate the correct daily dosage for a patient who weighs

- **a** 80 kg
- **b** 75 kg
- **c** 110 kg
- **d** 72.5 kg

2 A shop-owner uses this formula to calculate his monthly profits.

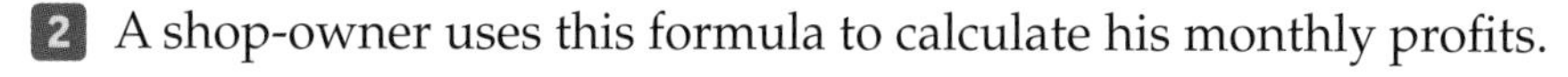

Profit = total takings minus total expenses

One month his total expenses are £4600.
Calculate his profit if his total takings are

- **a** £6500
- **b** £8000
- **c** £5000
- **d** £10 000

3 A Premiership football club uses this formula to calculate the number of stewards it needs on a match day.

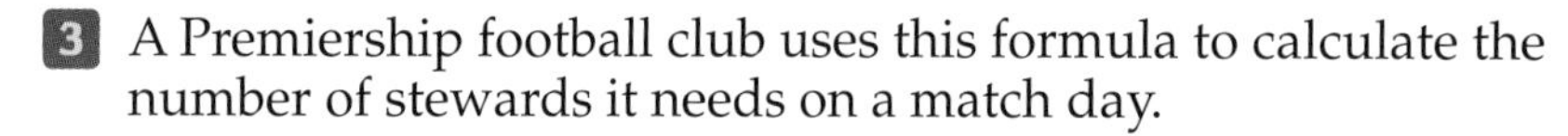

Number of stewards needed = number of fans divided by 200

Calculate the number of stewards needed if there are

- **a** 40 000 fans
- **b** 36 000 fans
- **c** 71 000 fans
- **d** 18 200 fans

9.2 Using formulae with two operations

Worked example: These are the cooking instructions on a leg of lamb.

Cooking time 40 minutes per kg plus an extra 20 minutes

A leg of lamb weighs 2.6 kg.
Calculate the cooking time for this leg of lamb.

Answer

40 × 2.6 = 104

104 + 20 = 124

This formula uses two operations. First you multiply by 40, then you add 20.

The cooking time is 124 minutes = 2 hours and 4 minutes. **(2 marks)**

1 The cost of a taxi ride is calculated using this formula.

Cost of taxi ride = £2 plus £1.40 per mile

Work out the total cost for a journey of

a 3 miles
b 7 miles
c 2.5 miles
d 6.2 miles

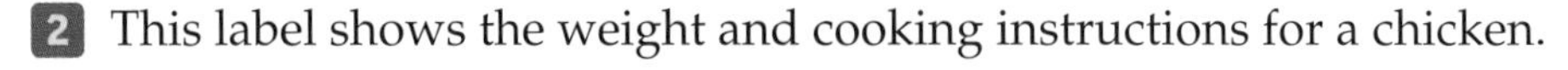

2 This label shows the weight and cooking instructions for a chicken.

a How much does this chicken cost?
b How long should this chicken be cooked for?
Give your answer in hours and minutes.

3 A cleaning company uses this formula to work out how much to pay its workers.

Weekly pay = £7.50 per hour of normal time plus £10 per hour of overtime

Calculate the weekly pay of an employee who works

a 35 hours of normal time and no overtime
b 35 hours of normal time and 3 hours of overtime
c 32 hours of normal time and 7 hours of overtime
d 33.5 hours of normal time and 4.5 hours of overtime

Data sheet

If you want to learn a skill or trade you can do an apprenticeship. Apprenticeships are offered by businesses and colleges. Apprentices earn at least £95 per week and get a qualification such as an NVQ or BTEC. There are over 190 different jobs to choose from.

Kara is looking for an apprenticeship on the internet. She has downloaded these details of two apprenticeship vacancies.

Tax facts

Most people who work in the UK pay income tax. This money is used to pay for public services like schools and the police.

You don't have to pay income tax if you earn less than £125 a week. If you earn £125 or more, you can work out roughly how much income tax you will have to pay each week using this formula.

> Subtract £125 from your weekly salary then divide by 5.

Tax is usually deducted by your employer. The amount of money you get after you have paid your tax is called your **net pay** or **take-home pay**.

Apprentice vehicle engineer

Location:	Oldbury
Type:	Employment with training
Age:	16+
Qualifications:	GCSE grade C or above in Maths and English
Hours:	Approx. 37 hours per week (varies)
Pay:	£4.20 per hour plus £20 per week
Holiday:	2 days for every month worked

Apprentice electrician

Location:	Walsall
Type:	Employment with training
Age:	16+
Qualifications:	At least 3 GCSEs grade C or above in Maths, English and Science
Hours:	Monday to Friday 8 am to 5 pm with 1 hour for lunch
Pay:	£4.80 per hour
Holiday:	20 days per year. Holiday entitlement will increase by 1 day per year for each year of experience (up to a maximum of 25 days per year).

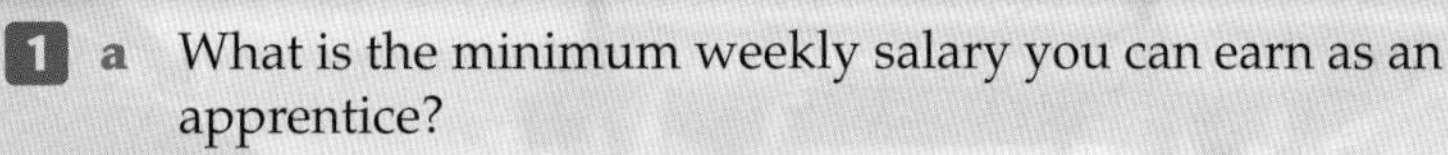

Questions

1 **a** What is the minimum weekly salary you can earn as an apprentice?

b If you earned this salary every week, how much would you earn in one year?

2 **a** How many hours would Kara work each week as an electrician?

> If you have an hour for lunch this does not usually count towards your working hours.

b Calculate Kara's weekly pay as an electrician.

c Kara's dad says that she would earn more money as a vehicle engineer. Is he correct? Show all of your working.

3 Kara is comparing the holiday allowances for the two apprenticeships.

a Kara has no experience. How many days of holiday would she get during her first year as an electrician?

b How many days of holiday would she get after two years as an electrician?

c If Kara worked a full year as a vehicle engineer how many days of holiday would she be entitled to?

4 Kara takes the apprenticeship as a vehicle engineer. She is planning her finances and wants to work out her take-home pay.

a Copy and complete this table showing take-home pay for different weekly salaries. Show all of your working.

Weekly salary	£100	£150	£200	£250	£300
Take-home pay		£145			

b In her first week Kara works for 35 hours. This is part of her pay slip for that week:

Employer: Oldbury Autos Ltd	Date: 17/6/2010	Name: Kara Ross	NI: SN 30 17 66 B
Payments		*Deductions*	
Wages: 35 hours @ £4.20 per hour plus £20		**PAYE income tax deducted** £15.20	

Has Kara been charged the correct amount of tax? Show all of your working.

Question key:

Q Open
Beginner
Improver
Secure pass

10 Perimeter, area and volume

Practise the maths

You need to calculate lengths and areas when you are doing DIY or gardening – for example, for working out how much weedkiller you need to cover a lawn.

10.1 Calculating the perimeter and area of a rectangle

Worked example: Work out

a the perimeter of this rectangle

b the area of this rectangle.

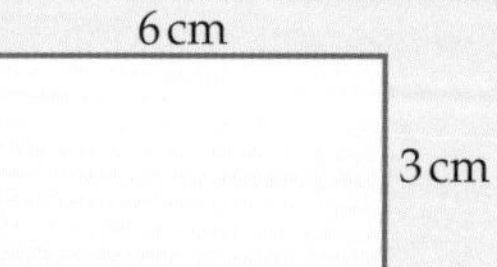

Answer

a

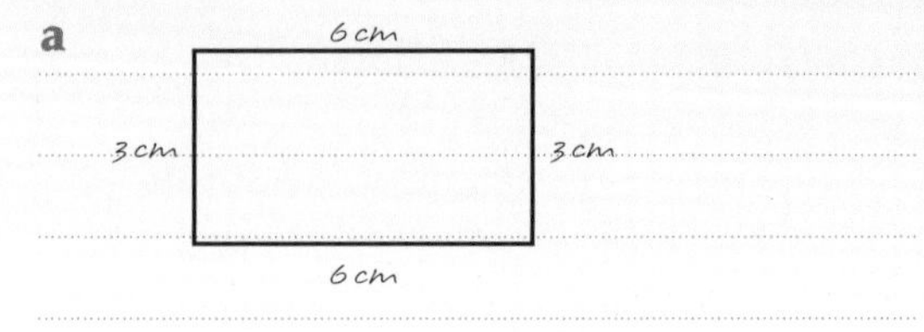

Perimeter

$= (6 + 3) \times 2 = 18$ cm **(1 mark)**

b *Area* $= 6 \times 3 = 18$ cm^2 **(1 mark)**

> **Perimeter** is the distance around the edge of a shape. Write in any missing dimensions to work out the perimeter.

> **Area of a rectangle** = length × width. The units of area are mm^2, cm^2 or m^2.

1 For each of these rectangles work out **i** the perimeter **ii** the area.

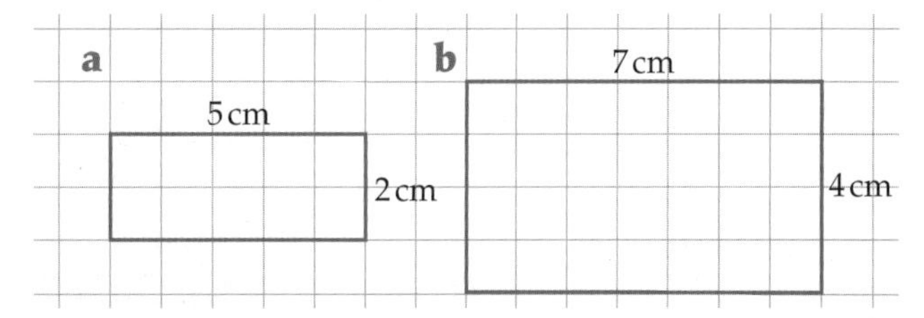

c 8 cm, 3 cm

d

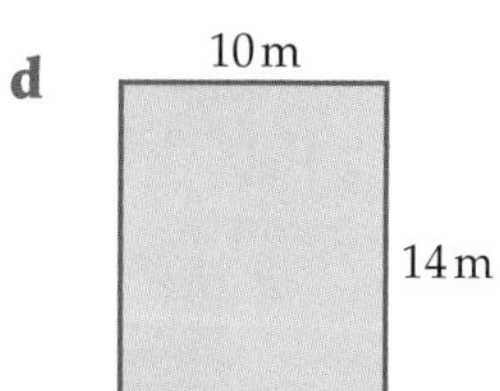

10.2 Calculating areas of shapes made from rectangles

Worked example: Work out the area of this shape.

8 cm, 5 cm, 11 cm, 7 cm

Answer

8 cm, 5 cm, 11 cm, 88 cm^2, 35 cm^2, 7 cm

Area $= 88 + 35 = 123$ cm^2 **(2 marks)**

> You can divide this shape into rectangles. Work out the area of each rectangle, then add the areas together.

1 Work out the area of this shape.

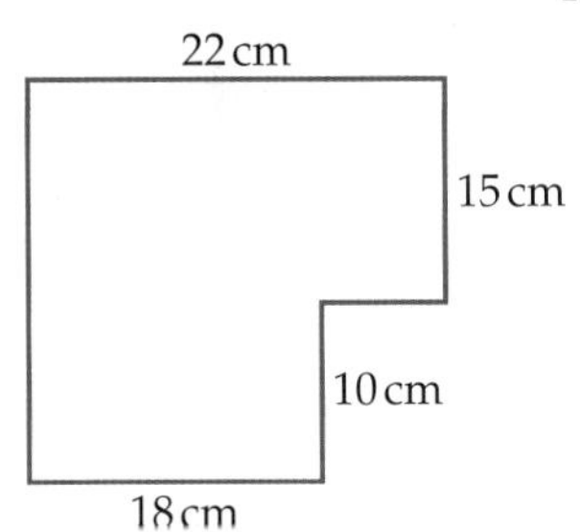

2 The diagram shows a plan of a wooden patio.

Maya wants to apply a coat of varnish to the patio.

She has enough varnish to cover 16 m² of wood.

Does she have enough varnish? Show your working.

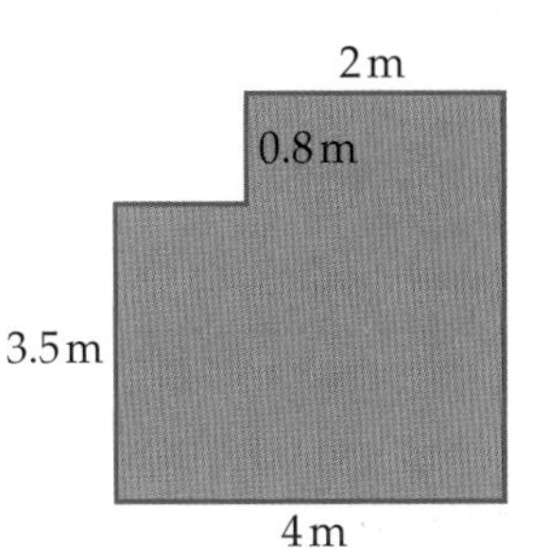

10.3 Calculating areas of triangles

Worked example:
Work out the area of this triangle.

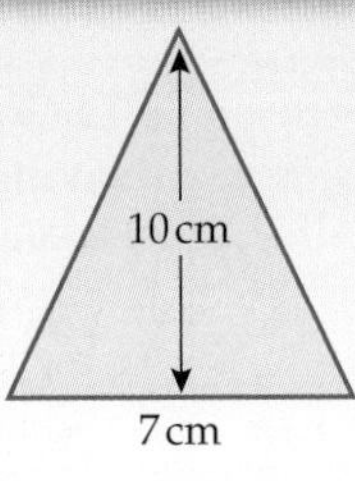

Answer

Area $= \frac{7 \times 10}{2} = \frac{70}{2} = 35\ cm^2$ *(2 marks)*

Area of a triangle $= \frac{\text{base} \times \text{height}}{2}$

1 Work out the area of each of these triangles.

a

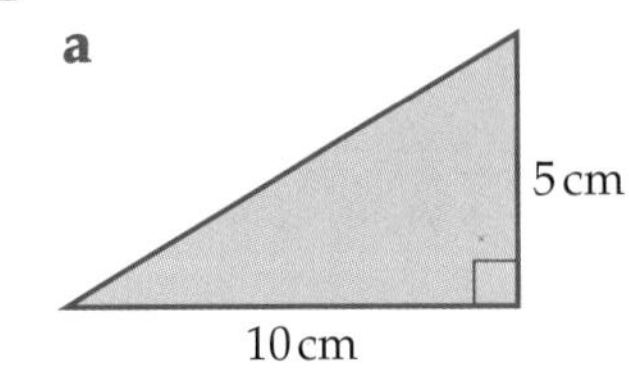

b

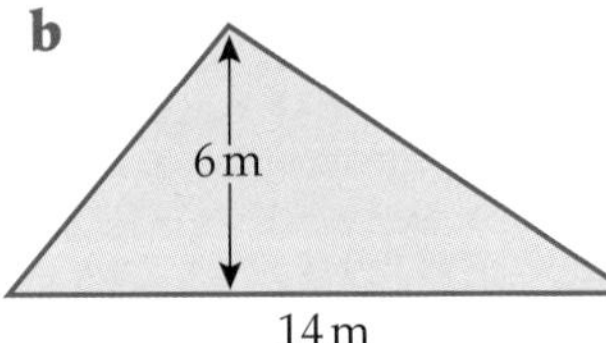

2 Calculate the total area of this shape.

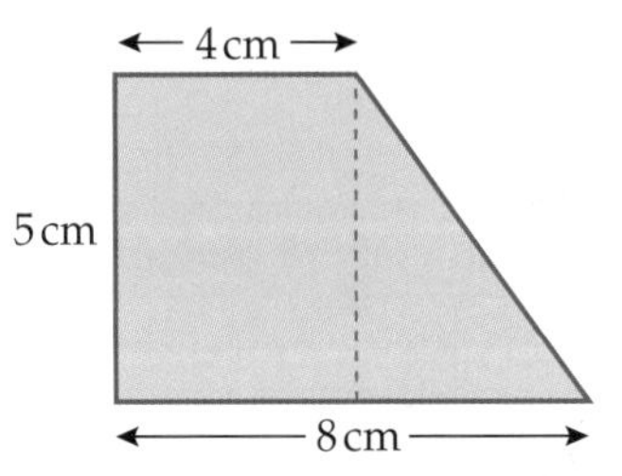

10.4 Calculating volumes of cuboids

Worked example:
Work out the volume of this cuboid.

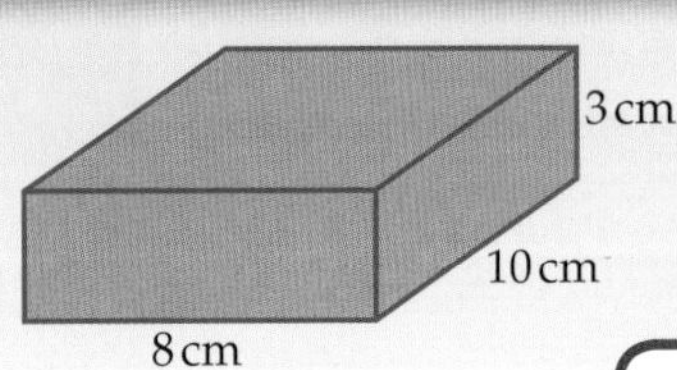

Answer

Volume $= 8 \times 10 \times 3 = 240\ cm^3$ *(2 marks)*

A **cuboid** is a 3-D shape. It has six rectangular faces.
Volume of a cuboid = length × width × height
The units of volume are mm³, cm³ or m³.

1 Work out the volume of each of these cuboids.

a

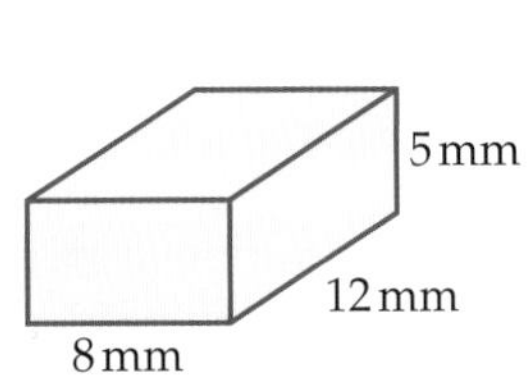

b

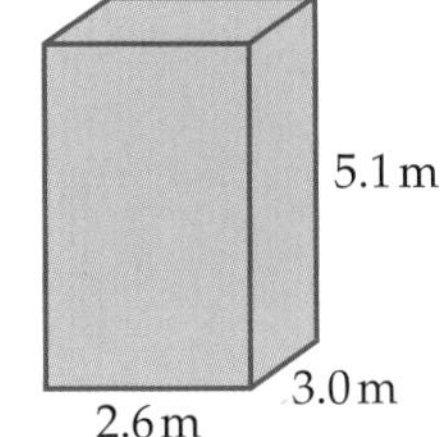

c

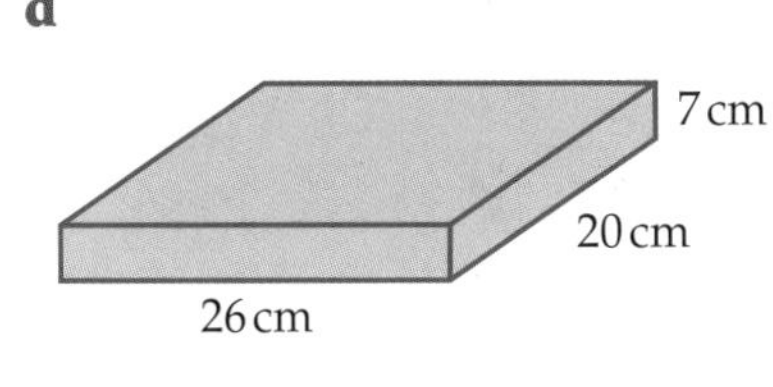

d

Data sheet

If you sell a house in the UK you must provide an Energy Performance Certificate. This tells a potential buyer about the energy efficiency and carbon dioxide emissions of the house. It also provides a list of practical suggestions for cutting fuel bills and carbon dioxide emissions.

Energy Performance Certificate: 22 Glenside Drive, Plymouth, PL2 6XF

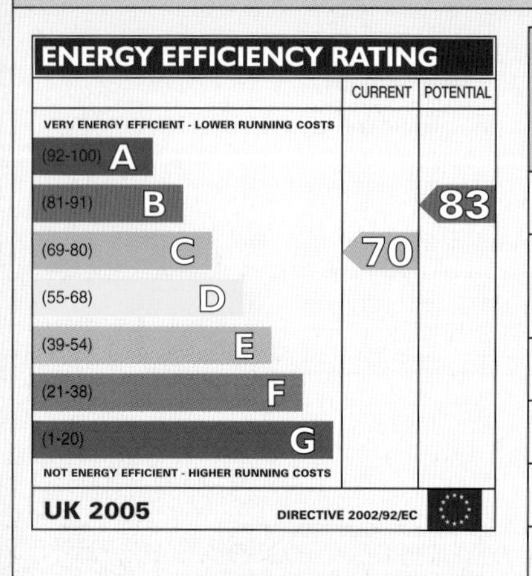

Suggested improvements	Typical savings per year
1. Loft insulation	£226
2. Insulating tape around windows	£18
3. Low energy lighting	£11
4. New boiler	£175
5. Double glazing	£190
Sub-total	£620

Estimated current carbon dioxide emissions and fuel costs	
Carbon dioxide emissions	13 tonnes per year
Lighting	£81 per year
Heating	£1473 per year
Hot water	£219 per year

What can we do about Global Warming?

Carbon dioxide is produced when electricity is created, or when oil and gas are burned.

The oil, gas and electricity we use in our homes generate 30% of all the carbon dioxide emissions in the UK.

If every household in the UK replaced one conventional lightbulb with an energy saving lightbulb we would save enough carbon dioxide each year to fill the Royal Albert Hall 2000 times.

Loft insulation

£14.50 per roll

1.1 m wide, 150 mm thick, 4 m long

Professional installation service: £140

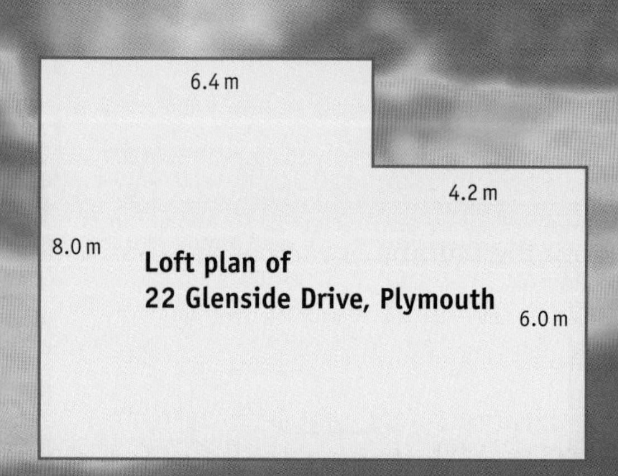

1 **a** What percentage of the carbon dioxide produced in the UK comes from energy use in homes?

b In 2009 the UK was responsible for producing 620 million tonnes of carbon dioxide. How many tonnes of carbon dioxide were produced by energy use in homes?

2 Jonah has just bought 22 Glenside Drive in Plymouth.

a What is the current energy efficiency rating of Jonah's house?

> Energy efficiency is rated from A to G.

b What is the potential energy efficiency rating of Jonah's house?

3 Jonah has a budget of £150 per month to pay for his lighting, heating and hot water. Is it enough? Show all your working.

> The table on the data sheet shows how much Jonah should expect to pay in one year.

4 Jonah's house needs 14 lightbulbs.
A pack of 3 low energy lightbulbs costs £8.99.

a How many packs of low energy lightbulbs will he need to buy?

b How much will they cost?

c How much money will Jonah save each year by making this improvement?

d Estimate the length of time it would take for this improvement to pay for itself.

5 Jonah decides to insulate his loft to save money on his heating bills.

a Work out the area of Jonah's loft.

b Each roll of loft insulation covers 4.4 m^2. How many rolls of loft insulation will Jonah need to buy? Use a diagram to explain how Jonah could lay out his loft insulation.

c Jonah decides to have his loft insulation professionally installed. Work out the total cost of insulating Jonah's loft.

d Estimate how many years it will take for this improvement to pay for itself.

6 This table shows the number and sizes of the windows in Jonah's house.

Window size	Small	Medium	Large
Dimensions	24 cm × 50 cm	48 cm × 50 cm	80 cm × 120 cm
Number in house	5	5	2

Jonah wants to put insulating tape around all of the windows in the house.

> **INSULATING TAPE £4.99 FOR 4 M**

Jonah says that the window insulation will pay for itself within 2 years.

Do you agree with Jonah's statement? Show all of your working.

Question key:
- Q Open
- > Beginner
- >> Improver
- >>> Secure pass

11 Geometric shapes

Practise the maths

Athletes use angles to work out the best way to throw a javelin. To get the longest throw you should release the javelin at an angle of 33° to the ground.

11.1 Lines of symmetry

Worked example: Draw in any lines of symmetry on these shapes.

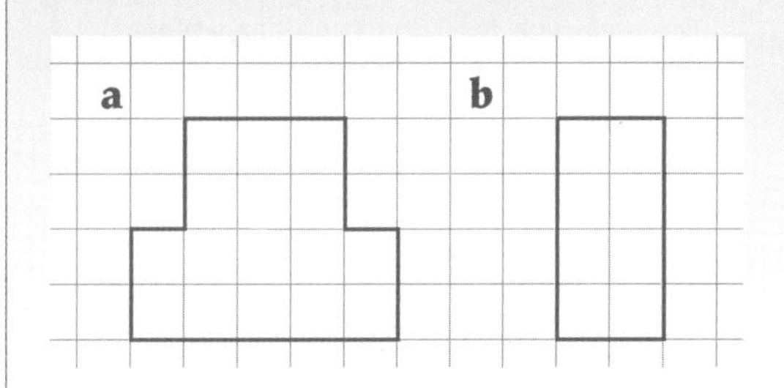

Answer

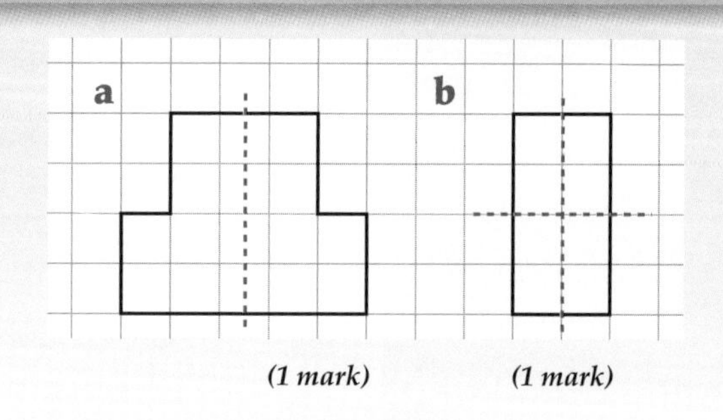

(1 mark) *(1 mark)*

A **line of symmetry** divides a shape in half.
Each half is a mirror image of the other half. You can use dotted lines to show lines of symmetry on a diagram.

1 Copy these shapes on squared paper. Use dotted lines to show all the lines of symmetry.

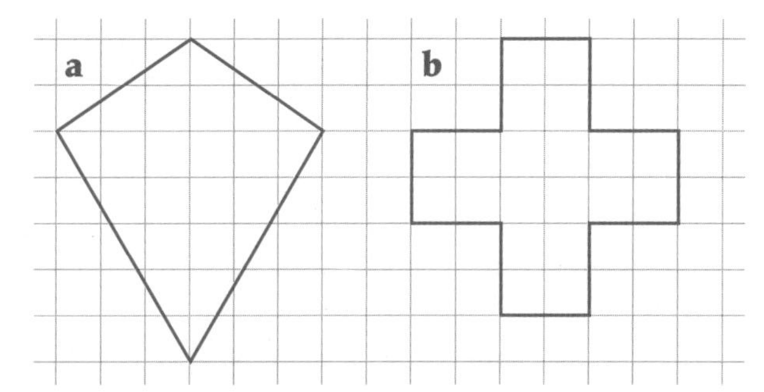

2 Copy this shape on squared paper. Complete the shape so that the dotted line is a line of symmetry.

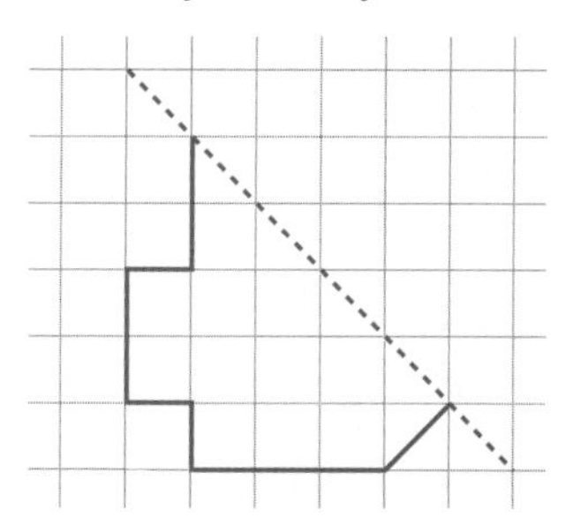

11.2 Measuring lines and angles

Worked example: Josh is using a protractor to measure this angle.

a Write down the size of the angle correct to the nearest degree.

b Is this angle acute or obtuse?

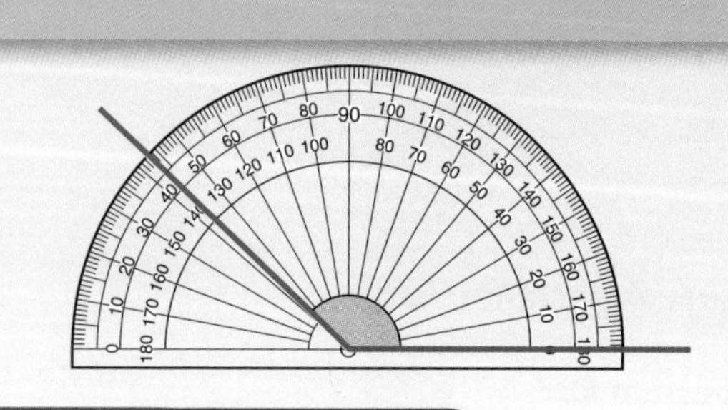

Answer

a *137°* *(1 mark)*

b *Obtuse* *(1 mark)*

Make sure you read the scale in the correct direction.

Angles smaller than 90° are **acute** angles.
Angles greater than 90° but less than 180° are **obtuse** angles.

1 Use a protractor to measure the size of each angle to the nearest degree.
Write down whether each angle is acute or obtuse.

a

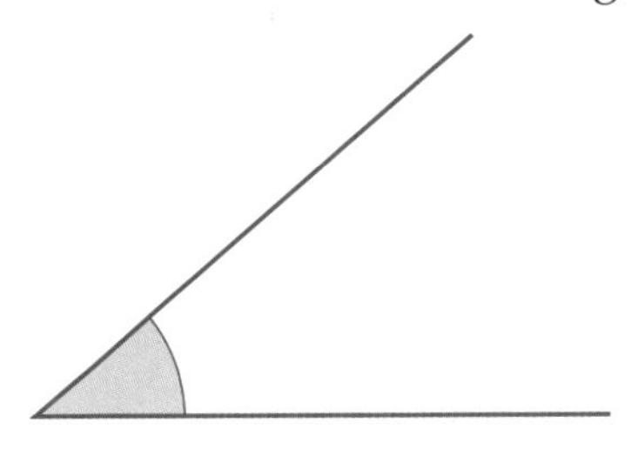

b

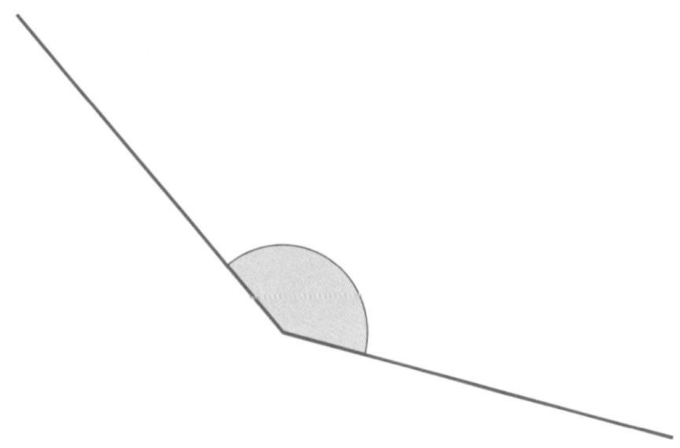

2 Measure the length of each line. Give your answers in cm to one decimal place.

a

b

3 Megan is measuring an angle with a protractor.
She says that its size is 25°.

a What mistake has Megan made?

b What is the size of the angle?

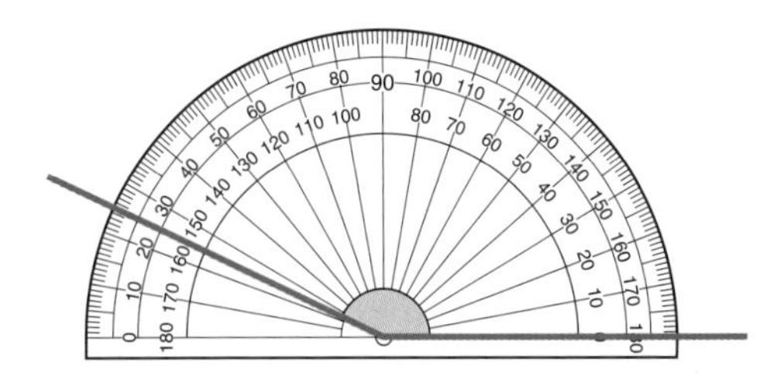

11.3 Drawing 2-D shapes

Worked example: Use a ruler and protractor to draw an angle of 62°.

Answer

Draw a line with your ruler. Place the centre of the protractor exactly at one end of the line. Draw a dot at 62°.

Use a ruler to join up the end of your line and the dot. Label the angle.

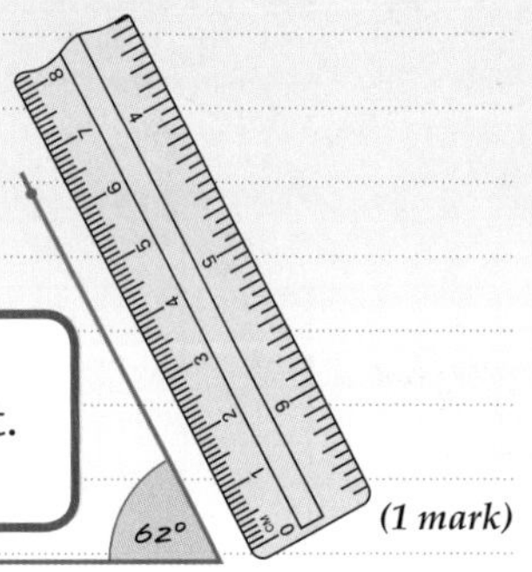

(1 mark)

1 Use a ruler and protractor to draw an angle of

a 40° **b** 77° **c** 118° **d** 96°

2 Use a ruler and protractor to make an accurate drawing of each of these shapes.

a

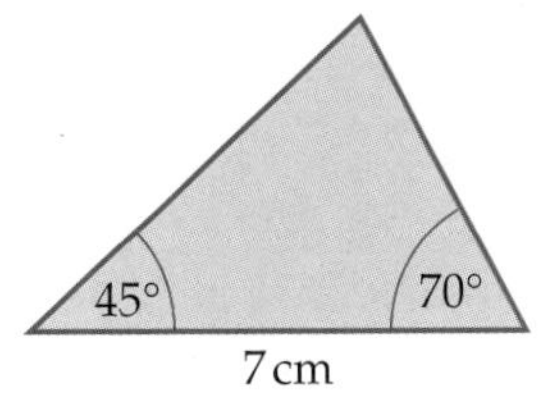

b

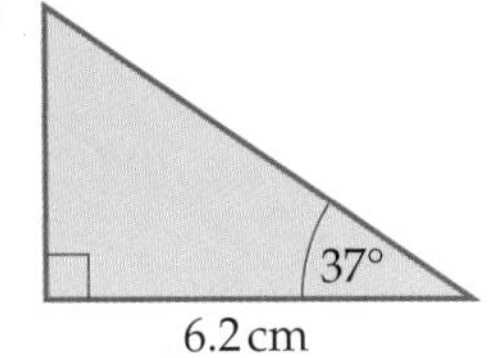

Data sheet

A vegetable garden is a good way to grow healthy food and save money.

The plan at the bottom of the page is a scale drawing of the vegetable garden at Woodshall High School.

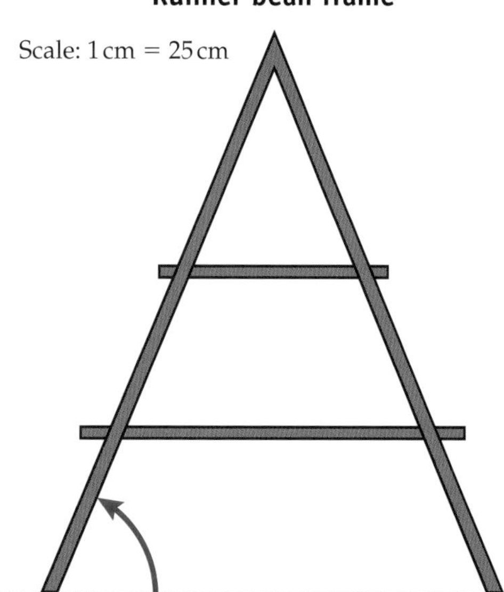

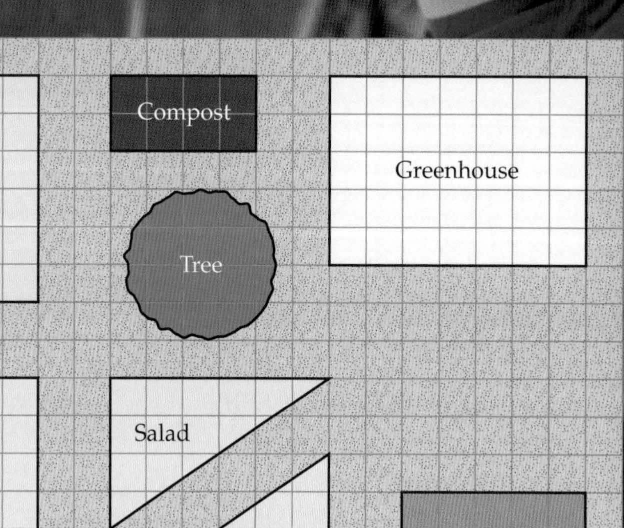

1 The plan of the school garden is drawn to a scale of 1 cm = 1 m.

a Calculate the width of the greenhouse in real life.

b Calculate the area of the carrot bed in real life.

c The school wants to install benches in the garden. The benches are 60 cm wide. How wide will the benches be on the scale drawing?

2 Wire rabbit fencing costs £8.50 per metre. Calculate the total cost of putting rabbit fencing around all six of the vegetable beds. Show all of your working.

3 David is planting runner beans on wooden frames.

a How many lines of symmetry does the frame have?

b Use a protractor to measure the size of the angle marked with an arrow on the diagram of the runner bean frame. Give your answer to the nearest degree.

c Estimate the total length of wood needed to make one runner bean frame.

4 Alison has designed wooden markers for each of the different vegetables. This diagram shows a blank marker. The diagram is **not** drawn to scale.

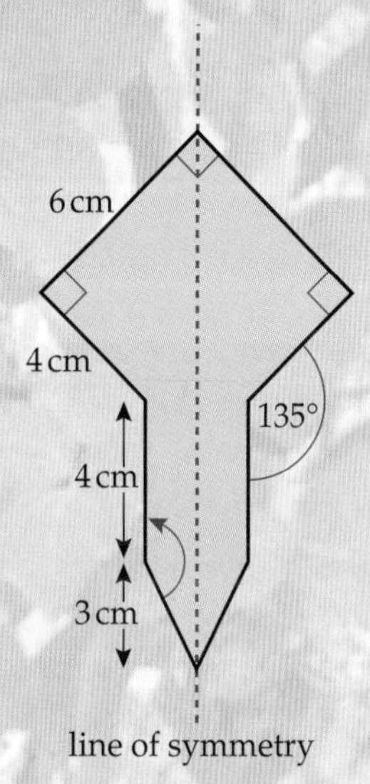

a Use a ruler and protractor to make an accurate drawing of the marker.

b Measure the size of the angle marked with an arrow. Give your answer to the nearest degree.

5 Sajid wants to design a flower bed to go between the greenhouse and the shed. Design a flower bed that will fit in this space. The flower bed needs a 50 cm wide path all around it and must have at least one line of symmetry. Make an accurate drawing of your flower bed using a scale of 1 cm = 50 cm. Mark any lines of symmetry on your drawing.

Question key:

- Q Open
- Beginner
- Improver
- Secure pass

Exam**Café**

Welcome to the Exam Café! This section of the book is here to guide you through the revision period.

First of all remember that examiners are looking to find out what you *can* do, not what you cannot do. The examiner is instructed ALWAYS to mark in a positive manner, so you just need to help him by always showing your working.

Basic exam info

Here's some basic information about the exam.

- For Level 1 you will sit a single Functionality paper in either January, March, June or November.
- The paper lasts 1 hour 30 minutes and has 60 marks.
- There will be between 3 and 5 questions all of which you should attempt.
- Calculators are allowed.
- Each individual question will be based on a single context, e.g. planning a visit to the London Eye, but will be split up into sub-parts.
- Up to 20% of marks are allocated to testing basic maths skills.

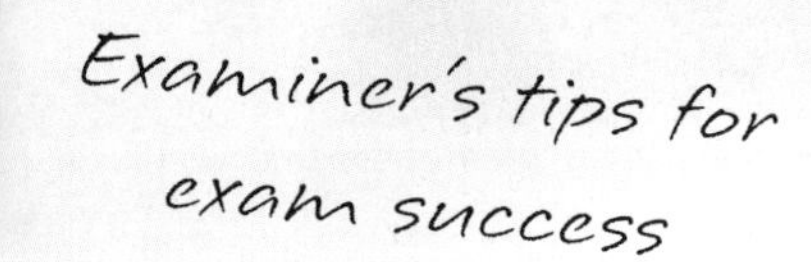

To be successful in Functional Maths, you have to know your basic maths! You need to be happy with FDPRP:

FRACTIONS, DECIMALS, PERCENTAGES, RATIO AND PROPORTION

Don't forget you will always have your calculator to help. But your calculator is not the answer to everything...

Don't depend on your calculator! Even though your calculator will help you to work out your answer, you must follow this rule of thumb:

ESTIMATE – CALCULATE – CHECK IT MATE!

ESTIMATE your answer first, then CALCULATE it using your calculator, then CHECK it using common sense.

WHY SHOULD I SUCCEED AT FUNCTIONAL MATHS

Maths for life
Functional Maths teaches you the maths you will need all the time throughout your life. You can use your maths skills to understand the world around you, or to succeed at work or in education.

MATHS AT HOME
- Choosing the cheapest car insurance – Calculations
- Putting up a wall bracket for a flat screen TV – Angles
- Working out what time to leave home to catch a certain train – Times and timetables

MATHS IN EDUCATION
- A-level Geography – Reading graphs
- Apprentice electrician – Using formulae
- Biology degree – Negative numbers

MATHS AT WORK
- Bookmaker – Probability
- Accounts clerk – Adding and subtracting decimals
- Architect – Plans and elevations
- Construction engineer – Using ratios

Show your working! Even though you are using a calculator...

YOU MUST SHOW ALL YOUR WORKING OUT

If you make an error, you can still win close to full marks by showing your working and showing the examiner that you understand the process and key steps. Remember: the examiner is always looking to award marks, not take them away.

Use this book! Every chapter starts with...

PRACTISE THE MATHS

This is where you can revise all your basic skills, and make sure you're ready to tackle Functional Maths questions set in context.

Data Sheets

All Functional Maths questions are set in everyday contexts. There are 3–5 questions in the exam, and each question comes with a data sheet, providing the material on which you will be tested.

The data sheets for 1 or 2 of the questions will be sent to your school or college 4 weeks before the exam, so you will have some time to prepare those questions in advance!

When you sit the exam you will be given a fresh **data book** that will contain *all* the data sheets for *all* the questions in the exam. This data book will include fresh copies of the data sheets that were sent in advance. You will **not** be allowed to bring your original copies into the exam with you.

Advanced release of data sheets - so what's the best way to prepare?

- A good way to do this is to work with your fellow students and your tutor and try to predict the sort of questions that could be asked from the data sheet.
- Many of the data sheets will have an example, but can you come up with another question that could be asked, and how would you answer it?
- Ask yourself the question: "What is the data telling me?"
- By looking at the information given you can familiarise yourself with the context or ask your tutor to explain it to you if you are unsure.
- Your copy of the data sheet is there to be annotated, but you will not be able to take it into the exam.
- In the exam, you will receive a clean copy as well as further data sheets for the other questions, so not everything can be prepared in advance.

DON'T SPEND TOO MUCH TIME ON THESE SHEETS

– just an hour or two on each one.

Instead, organise yourself a revision schedule so that you

o revise all the worked examples in this book and the functional skills pages that are dedicated to your level

o practise exam questions: you will find a Practice Paper at the end of the Exam Café, and your teacher will be able to provide you with more.

What to do before and during your exam

DO	DON'T
Make sure you have pens, pencils, a ruler, a protractor, a compass and a calculator in a see-through bag or pencil case	Don't forget to check that your pens work and your pencils are sharp
Write legibly in a black or blue pen	Don't write in pencil unless you are drawing a diagram or constructing lines or angles
Make sure you have a scientific calculator and know how to use it	Don't forget your calculator or use one you are not familiar with
Think about what you are going to write before you start and keep to the point	Don't "waffle" or write answers if your working doesn't support them
Make sure you attempt each question in order	Don't leave any questions out
Show all your working out in step-by-step, logical calculations	Don't just write an answer without any working or guess answers
Put a line through any working you don't want to be marked	Don't use correction fluid or scribble out answers
Stick to your first answer unless you are sure your original working is wrong	Don't cross out working without replacing it
Make sure your graphs are the right size and you have used a suitable scale on your axes	Don't make your graph too large or too small for the graph paper given
Make notes and underline or highlight the important words in a question	Don't ignore, miss out or misread any of the words in the question
Make sure you have answered every part of the question in the correct order	Don't forget that you may have to write a final sentence to fully answer the question
Read all the information on the data sheet and keep referring back to it	Don't just scan over the data sheet information
Make use of any tables, graphs and charts you are given and annotate them to help you with your calculations	Don't leave a graph or table blank because you are worried about writing something incorrect
Make sure that your answers look sensible	Don't write an answer from your calculator without checking that it makes sense
Go back over your answers if you have any time left at the end of the exam	Don't leave the exam early

Exam Glossary

Functional Maths exams use a particular language. Study the terms below so that you're ready when you sit your exam.

If the question says...	Then you have to...
Analyse	Use the right mathematical methods to carry out step-by-step calculations and check your answer. You might need to use information from the data sheet for this question.
Discuss	Look at the data given to you and show working to support your conclusions. You might need to use information from the data sheet for this question.
Interpret	Use words and diagrams to show how your calculations relate to the question.
Justify your answer	Use words and/or diagrams to support your answer.
Write down	Write down the answer. You don't need to show working for this question, though you might want to anyway!
Give a counter-example	Give an example that shows that the statement in the question is not true. This example could be a value or a diagram.
Make an accurate drawing/use a ruler and compasses	Use a pencil, ruler, protractor and compasses as necessary to construct and measure lines, angles, arcs etc.
Not drawn accurately/ not to scale	Use calculations to find missing lengths and angles. Make sure that you don't measure the lengths and angles on the diagram to find your answer.
Explain	Use words and calculations to explain how you found your answer.
You MUST show your working	Show all of your calculations. If you just give an answer to this question you won't get any marks.
Estimate	Round the numbers given to 1 significant figure. You can use the rounded numbers in your calculations to estimate the answer.
Show	Write down any working or diagrams that are necessary to reach the answer or value given in the question.
Prove	Use logical steps to demonstrate how you have reached your answer. If you use any mathematical facts in your working you must write them down.
Work out	Work out a calculation either mentally, using a written method, or with a calculator.

Calculate	Use a calculator or a written method to find an answer.
Hence	Use the previous answer to help you find a solution.
Hence, or otherwise	Either use the previous answer or show new working to find a solution.
Measure	Use a ruler or protractor to measure a length or angle correct to a given degree of accuracy.
Use the graph	Use the graph given to read off your answer. You can write on the graph to show your working if the graph is in the question. If the graph is only provided on the data sheet, you can still write on it, but remember that the examiner will not see this.
Describe fully	Write down the type of transformation and all the information needed to define it. • Reflection: describe where the mirror line is. • Rotation: write down the centre of rotation, the angle of rotation and its direction. • Translation: write the translation as a column vector. • Enlargement: write down the centre of enlargement and the scale factor of the enlargement.
Comment on	Look at the information in the question and on the data sheet then use calculations and words to write your answer.

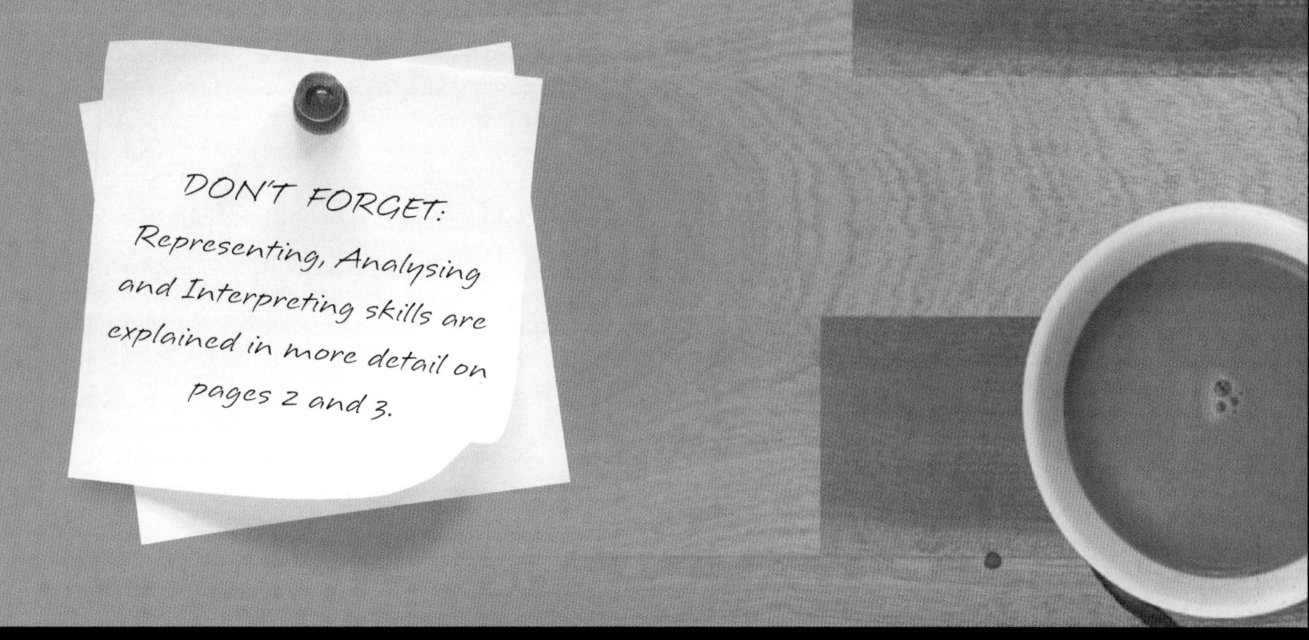

ExamCafé

Exam question

In a sale a settee is advertised with 60% off the original price of £800.
For one day only, the settee is offered with an extra 20% off the sale price.
Show that the total saving is 68% off the original price. **[3 marks]**

Student answers

A

$\frac{68}{100} \times 800 = 544$

This student has made the very common mistake of finding 68% of the original price rather than 68% off the original price. The student has also made no effort to compare the saving of 68% to a saving of 60% followed by a saving of 20%. The method of calculating the percentage is correct, so 1 mark is awarded.

B

$60\% + 20\% = 80\%$ off in total

$\frac{80}{100} \times 800 = 640$

$\frac{68}{100} \times 800 = 544$

Therefore the saving is not correct

The student's working is incorrect. Adding percentages when there is a repeated percentage change is a very common error. 1 mark is awarded for a correct calculation of 68% of 800.

C

$60\% + 20\% = 80\%$ off

$0.2 \times 800 = 160$ total cost

The student has incorrectly added the percentages. No marks are awarded.

D

60% off $800 = 480 - 800 = 320$

80% off $800 = 640 - 800 = 160$

$\frac{384}{800} \times 100 = 48$

The student has written the subtractions in the wrong order. However, the intended meaning is clear so 1 mark is awarded for correctly calculating a 60% reduction of 800.

E

$800 - 480 = 320$

$\frac{20}{100} \times 320 = 64$

$320 - 64 = 256$

The student has correctly worked out the final cost of the settee. This gains 2 marks. The candidate has made no attempt to compare their answer with the 68% reduction so they are not awarded the third mark.

Improved answers

This type of percentage question will frequently appear on Functional Maths papers. There will always be more than one method for calculating the correct answer so there are many ways to gain marks.

Using a correct method for calculating a 60% reduction (not finding 60%) of £800 gains the first mark. If £320 is written anywhere in the answer this will be enough to gain this mark.

The second mark is awarded for calculating a 20% reduction of the sale price. If the sale price (£320) has been incorrectly calculated but the 20% reduction has been correctly calculated this mark will still be awarded.

The final mark is very rarely achieved. It is awarded for showing that the final sale price is the same as a 68% reduction. Any candidate who calculated £256 using two different methods would probably achieve maximum marks.

Exam question

A bus runs from Glasgow to Leeds once a day except on New Year's Day and Christmas Day.

In 2007, the bus was on time for exactly two-thirds of its journeys.

Richard says that the bus was on time on exactly 250 days in 2007.

Is Richard correct?

You must show your working. **[3 marks]**

Student answers

A No!

The student has written the correct answer. However there is no working out shown so no marks are awarded.

B $\frac{250}{360} = 0.7$

The student has attempted to write the 250 days as a fraction of the entire year. However they have used the wrong denominator (360 instead of 363). The student would need to show more working to compare their fraction with $\frac{2}{3}$, and attempt to answer the question to gain any marks.

C $366 - 2 = 364$ and Yes

366 is not the number of days in 2007. No further working is shown so no marks are awarded.

D $\frac{364}{3} = 121.3$

$121.3 \times 2 = 243$,

Therefore Richard is wrong

The student has used an incorrect number of days (364 rather than 363) but has used a correct method to calculate $\frac{2}{3}$ of their value. 1 mark is awarded for this correct method and 1 follow-through mark is awarded for an attempt to compare their answer to 250 to answer the question.

E $363 - 250 = 113$, $113 \times 3 = 339$, No

The student is awarded 1 mark for using the correct total number of days (363), 1 method mark for a relevant calculation, and 1 accuracy mark from their working.

F $250 \times \frac{3}{2} = 375$ and No

The student has carried out a relevant calculation accurately, and has answered the question. All 3 marks are awarded.

DON'T FORGET:

Always show your working!

AQA Functional Mathematics

Level 1 Practice Paper - 1 hour 30 minutes 60 marks

1 A holiday in Barcelona

Data sheet

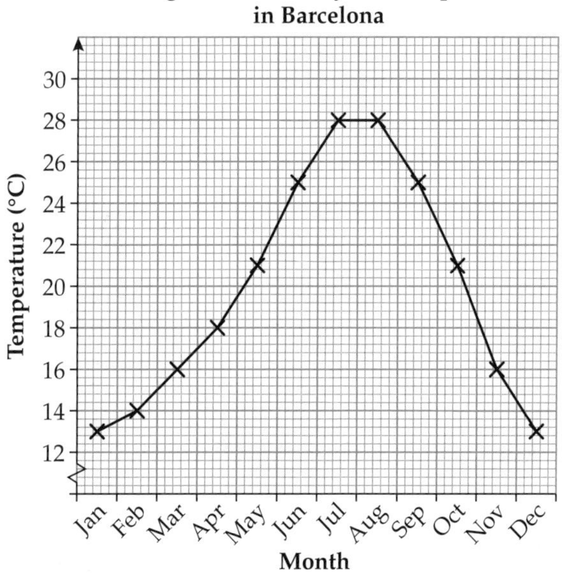

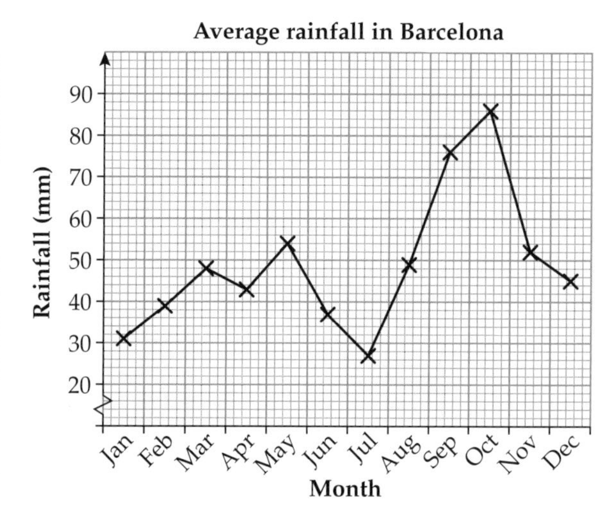

Prices of holidays in Barcelona (per person)			
Month	**Hotel Mar-menor**	**Hotel San Pedro**	**Hotel Sucina**
Jan–Feb	£648	£688	£708
Mar–Apr	£659	£699	£708
May–Jun	£735	£785	£815
Jul–Aug	£689	£769	£799
Sep–Oct	£675	£699	£789
Nov–Dec	£655	£689	£765
Customer satisfaction rating	* * *	* * * *	* * *

Price per person includes:
Return flights, seven nights' bed, breakfast and evening meal.

Holiday extras:
Room with a sea view, £98 per room per week.
Lunch, £7.50 per person per day.

Questions

1 a Mrs Slater plans a holiday to Barcelona.
She finds two graphs showing information about Barcelona.
The first shows the maximum average daytime temperatures in Barcelona.
The second shows the average rainfall in Barcelona.

Mrs Slater wants to go on holiday when the maximum daytime temperature is less than 26 °C. She wants to do a lot of sightseeing, so she doesn't want too much rain.

During which month do you suggest Mrs Slater goes to Barcelona?
You must give reasons for your answer. (*4 marks*)

b Mr Williams also plans a holiday to Barcelona.
He wants to book a double room with a sea view for himself and his wife.
He wants a room for 7 nights beginning on 26 May.
He can spend a maximum of £1800 on hotel accommodation, flights and lunch on six days.
Which hotel do you suggest Mr Williams should book?
You must show all your calculations and give reasons for your answer. (*6 marks*)

c Mr and Mrs Williams can catch a flight that leaves Bristol airport at 10:35 or 17:40.
It takes them between 30 and 45 minutes to drive from their home to Bristol airport.
When they arrive at the airport they must go to the check-in desk.
The check-in desk opens 2 hours before their flight leaves and closes 30 minutes before their flight leaves. It usually takes between 5 and 20 minutes to check-in and between 15 and 30 minutes to go from the check-in desk to where they board the plane.
Passengers usually start to board the plane 10 minutes before the plane is due to leave.

Write a plan showing their time schedule. You must include in your plan the time they should leave their home to travel to Bristol airport.
You must give a reason for choosing either the 10:35 or 17:40 flight. (*6 marks*)

2 Beachwatch

Data sheet

Beachwatch is a litter survey carried out on some of the beaches in the UK. The table shows some of the results of the Beachwatch survey in 2009.

Country	Number of beaches surveyed	Total number of items of litter	Total length of beach surveyed (km)
Northern Ireland	13	11 893	6.7
Scotland	61	46 763	24.5
Wales	73	59 226	19.2
England	217	206 959	116.4

Questions

2 a Sandra says, 'England has the biggest problem with beach litter'.

i Work out the mean number of items of litter per kilometre of beach surveyed for each country. Give your answers to the nearest whole number. *(4 marks)*

ii Comment on Sandra's statement about the country with the biggest litter problem. *(2 marks)*

b The pie charts show the sources of all the litter found on the beaches in 1995 and 2009.

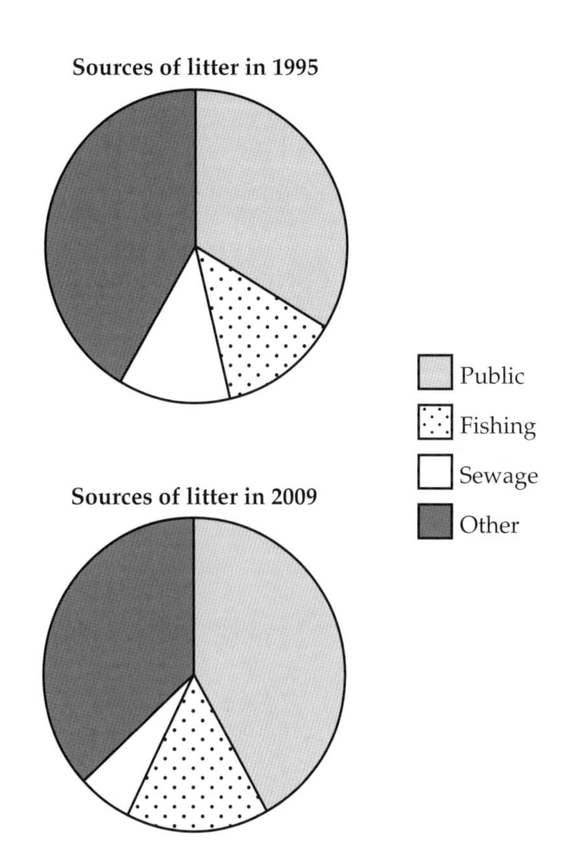

i Sandra says, 'In 2009, most of the litter found on the beaches was left by the public'.
Explain how you can tell from the pie chart that Sandra is correct. (*1 mark*)

ii Sandra says, 'The number of items of sewage litter found on the beaches has halved from 1995 to 2009'.
Is Sandra's statement correct? You must explain your answer. (*2 marks*)

c Sandra wants to compare the percentage of litter that comes from fishing in Northern Ireland, Scotland, Wales and England.
The table shows the percentages.

Country	**Northern Ireland**	**Scotland**	**Wales**	**England**
Percentage of litter that comes from fishing	8%	7%	27%	14%

Draw a suitable chart or diagram to display this information. (*5 marks*)

3 Horse feed

Data sheet

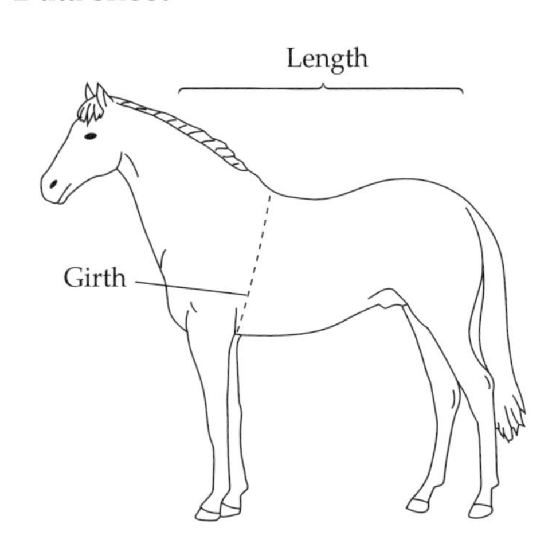

Bodyweight calculator

Girth (cm): 120, 130, 140, 150, 160, 170, 180, 190, 200, 210, 220, 230, 240

Weight (kg): 125, 150, 175, 200, 225, 250, 275, 300, 325, 350, 400, 450, 500, 550, 600, 650, 700, 750, 800, 850, 900, 950, 1000

Length (cm): 75, 80, 85, 90, 95, 100, 105, 110, 115, 120, 125, 130, 135, 140, 145, 150

To work out an estimate of the weight of a horse:

1 Measure the length of the horse in cm.

2 Measure the girth of the horse in cm.

3 Draw a line between the two measurements on the bodyweight calculator. Where your line crosses the weight scale is an estimate of the weight of the horse.

Questions

3 a Cheryl wants to work out an estimate of the weight of her horse. The tape measures show the length and the girth of her horse.

Length

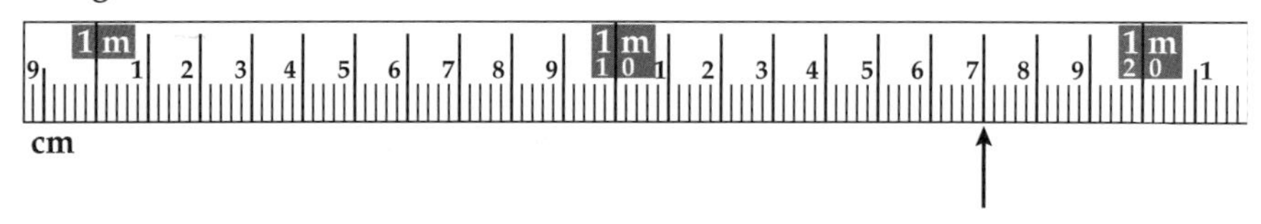

Girth

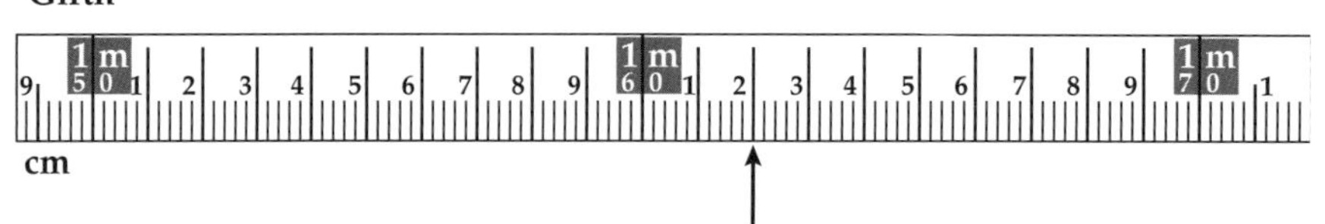

Cheryl says that the bodyweight calculator shows that her horse weighs 330 kg. Explain why she must be wrong. *(4 marks)*

b A formula to work out the amount of food a horse needs each day is:

daily amount of food in kg = weight of horse in kg ÷ 40

Horses are fed a mixture of hay and hard feed such as oats.
The amount of hay and hard feed they are fed depends on the horse's level of exercise.
The table shows the percentages of the daily amounts of hay and hard feed that are needed.

Level of exercise	Hay	Hard feed
Light	80%	20%
Medium	70%	30%
Hard	60%	40%

Cheryl says that her horse usually has a light to medium level of exercise.
Work out an estimate of the amount of hay and hard feed that Cheryl should feed her horse each day. *(8 marks)*

c This is the horse feed price list from the company that Cheryl uses.

Item	Price per item
Cool mix (20 kg)	£9.10
Speedibeet (20 kg)	£8.90
Leisure mix (20 kg)	£9.20
Pasture mix (20 kg)	£8.40
Pasture cubes (20 kg)	£7.70
Staypower cubes (20 kg)	£9.10
Apple chaff (15 kg)	£5.80
Molassed chaff (15 kg)	£5.50

Delivery charges

Total weight of order (kg)	Delivery charge
1–30	£5.71
31–50	£11.42
51–70	£17.13
71–99	£22.84

Free delivery on orders over £40 **and** total weight of order less than 30 kg.

This week's special offer!

Buy two bags of apple chaff and get the second half price.

Cheryl orders 1 bag of speedibeet, 1 bag of pasture cubes and 2 bags of apple chaff.
Show that the total cost of the horse feed and delivery comes to just over £40.
You must show all your working and check your answer using estimation. *(8 marks)*

4 Delivery van

Data sheet

The diagram shows the positions of five towns and how main roads link them.
It also shows the distances, in miles, between the towns.

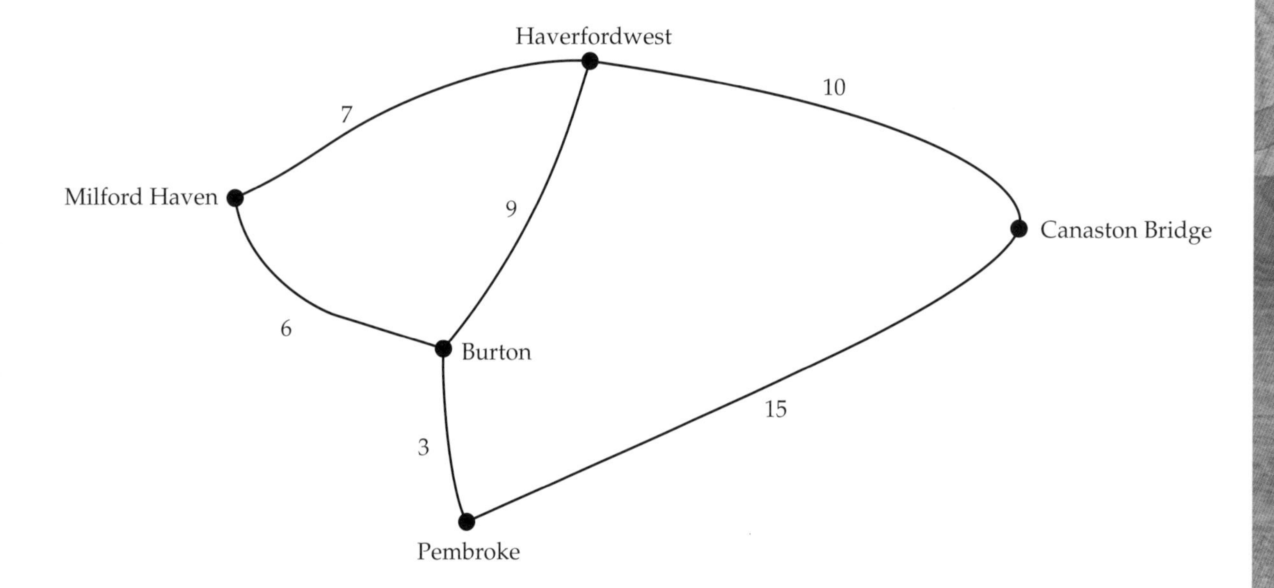

Questions

4 a Barry drives a delivery van. The company he works for is in Burton. It doesn't matter in which order he makes his deliveries, but he must start and finish in Burton.

One day he has to make a delivery in Haverfordwest, Pembroke and Milford Haven.
Write a plan for two different routes that he could take.
Which route do you think is the best route for Barry to take?
You must give a reason for your answer and show all your workings. (*5 marks*)

b On another day, the road from Burton to Haverfordwest is closed.
Barry has to make a delivery in Canaston Bridge and Milford Haven.
Which route do you suggest he takes?
You must give a reason for your answer and show all your workings. (*5 marks*)

Introduction to Level 2

Functional Maths is all about using maths in real-life situations. It involves taking mathematical skills you have already learned and applying them in everyday contexts, as well as in new and challenging situations.

This means Functional Maths is about more than just numbers – **it is about numbers and words**. To answer functional questions successfully, you will often need to provide written answers, conclusions and explanations. There's nothing to worry about – these are skills you use all the time in your English lessons. The purpose of this book is to show you how to use these skills in maths as well.

In order to use maths effectively, you must be able to understand information given to you, use it and interpret the results of your calculations. These are the functional maths skills that you need to develop and are called the **Process skills**.

The Process skills are divided into three sections: **Representing**, **Analysing** and **Interpreting**.

Process skills - Representing, Analysing and Interpreting

What is Representing?	What is Analysing?	What is Interpreting?
Representing means that you need to look at a problem, and work out what you need to do to solve it. You need to decide • what pieces of the information you need • what skills you need to use to solve the problem, e.g. addition, subtraction, multiplication etc • in what order you need to work things out.	Analysing means that you need to be able to use the skills you have decided you need to use, and work out the answers. You need to • work things out in the order you have already decided • use the mathematical skills already decided on to work out answers • check your calculations at every stage to make sure the answers make sense • write down all the calculations that you do and explain why you are doing them.	Interpreting means that you need to be able to use the answers to the calculations you have done. You need to explain what the answers mean and how they relate to the problem you've been asked to solve. You need to • explain what it is that you have worked out • explain how your answers relate to the problem you've been asked to solve • draw conclusions based on the comparisons that you have made.

Top tips – use DICE

- Decide - what is your plan? Write down every step.
- Information - write down everything that you use from the data sheet.
- Calculations - write them all down.
- Explain - what do your answers mean?

Process skills – Sample question and answer

Take a look at the question and model answer below.

It shows where the three different **Process skills** are used.

Here is the data sheet that provides all the information you need to answer the question.

Data sheet:

Mixes for concrete

Concrete	sand : cement
general building (above ground)	5 : 1
general building (below ground)	3 : 1
internal walls	8 : 1
paving	3 : 1

Notes:

- For general building and internal walls, use soft sand (or builder's sand).
- For foundations and paving, use sharp sand.
- Sand and cement are sold in 25 kg bags.

Question:

George is a builder. He is doing some renovations on a house.

He mixes his own concrete using a sand and cement mix.

He estimates that he will need 180 kg of concrete for internal walls and 330 kg for general building above ground.

Work out the amount of sand and cement that George needs to buy.

Model answer showing the three different Process skills:

Representing:

1. *work out the amount of sand and cement he needs for the internal walls using +, ÷ and x*
2. *work out the amount of sand and cement he needs for the general building above ground using +, ÷ and x*
3. *work out the total amount of sand and cement he needs*
4. *work out the amount of sand and cement he needs to buy*

Analysing:

1. *For internal walls, sand : cement ratio is 8 : 1*
   8 + 1 = 9, 180 kg ÷ 9 = 20 kg
   sand = 8 x 20 kg = 160 kg, cement = 1 x 20 kg = 20 kg
2. *For general building above ground, sand : cement ratio is 5 : 1*
   5 + 1 = 6, 330 kg ÷ 6 = 55 kg
   sand = 5 x 55 kg = 275 kg, cement = 1 x 55 kg = 55 kg
3. *total sand = 160 + 275 = 435 kg*
   total cement = 20 + 55 = 75 kg

Interpreting:

4. *Sand and cement are sold in 25 kg bags.*
   435 ÷ 25 = 17.4, 75 ÷ 25 = 3
   George needs to buy 18 bags of sand and 3 bags of cement.

Why should I show my working?

In a Functional Maths exam it really matters if you **show your working**. Of course, it's best if you answer every question correctly, but that's not always possible. The point is, a wrong answer with no working will get no marks, while a wrong answer with lots of working can get nearly full marks.

Below is an example of part of a Level 2-style data sheet and question.

After the question you will see a model answer.

Then you will see three different solutions given by candidates.

All the candidates have the same wrong answer, but they each score a different number of marks. The examiner's comments show where the marks are given to each candidate.

Data sheet

Trips away from home within Scotland in 2007 and 2008

Trip information	2007	2008
Number of trips	12.8 million	11.8 million
Number of nights away	48 million	45 million
Total amount spent on trips	£2856 million	£2871 million

Question:

What is the difference in the mean amount spent per night on trips made away from home within Scotland in 2007 and 2008? **(3 marks)**

Model answer:

Mean amount spent per night in 2007 = £2856 million ÷ 48 million = £59.50

Mean amount spent per night in 2008 = £2871 million ÷ 45 million = £63.80

Difference in the mean amount spent = £63.80 − £59.50

= £4.30

Candidates' answers		Examiner's comments
£7.90	0 marks	Candidate has shown no working, the answer is wrong, I award no marks.
£2856 million ÷ 48 million = £55.90 *difference = £7.90*	1 mark	Candidate gets the first mark for showing 2856 ÷ 48, even though the answer is wrong. Candidate hasn't shown where the £7.90 comes from, so I award no more marks.
£2856 million ÷ 48 million = £55.90 *£2871 million ÷ 45 million = £63.80* *£63.80 − £55.90 = £7.90* *difference = £7.90*	2 marks	Candidate gets the first mark for showing 2856 ÷ 48, even though the answer is wrong. Candidate gets the second mark for showing that they are subtracting their two answers. Candidate doesn't get the last mark as the answer is wrong.

How should I answer an open question?

An **open question** is one where there may not be a single correct answer. There may be many different correct answers.

When you answer an open question, think **DICE** – see page 66.

Below is a Level 2-style question with two sample answers.

Data sheet

Multi-fuel stove information

Model number:	B47	B52	K87	K89	M45
Height (cm)	62	56	61	23	58
Width (cm)	65	66	57	44	70
Depth (cm)	39	42	34	34	46
Weight (kg)	125	95	92	72	135
Kilowatt rating	12	7	8	6	16
Price	£350	£320	£290	£235	£425
Customer rating	****	*	**	****	*****

Question:

Q Sham decides to buy a multi-fuel stove to heat his lounge.

He has a rectangular lounge measuring 5.2 m wide by 6.8 m long by 2.9 m high.

Sham must buy a stove with a kilowatt rating greater than the kilowatts needed to heat his lounge.

Sham works out the kilowatts needed to heat his lounge using this formula:

$$\text{Kilowatts needed to heat a room} = \frac{\text{volume of room (m}^3\text{)}}{14}$$

Sham needs to buy a stove that has a depth less than 40 cm. The height and width of the stove do not matter.

Which model of stove do you recommend Sham buys?

Candidates' answers:

Candidate A

Plan:
1. *work out the volume of Sham's lounge*
2. *work out the kilowatts needed to heat the lounge*
3. *decide which stoves Sham cannot buy*
4. *decide which stove Sham should buy and explain my answer*

1. *volume = 5.2 x 6.8 x 2.9 = 102.544 m³*
2. *kilowatts needed = 102.544 ÷ 14 = 7.32...*
3. *Sham needs a stove with a kilowatt rating of 8 or more, so he could buy the B47, the K87 or the M45. However, the depth of the M45 is greater than 40 cm, so he can't buy this one.*
4. *I think Sham should buy the B47 as it has a much bigger kilowatt rating than he needs which means it will keep the room really warm. Also it has a very good customer satisfaction rating.*

Candidate B

Plan:
1. *work out which stoves have a depth greater than 40 cm*
2. *work out the kilowatts needed to heat the lounge*
3. *decide which stoves Sham could buy*
4. *decide which stove Sham should buy and explain my answer*

1. *The B52 and the M45 have a depth greater than 40 cm so Sham cannot have either of these.*
2. *kilowatts needed =* $\frac{5.2 \times 6.8 \times 2.9}{14} = 7.32...$
3. *Sham could buy either the B47 or K87.*
4. *I think Sham should buy the K87 as it has a kilowatt rating just greater than he needs for his lounge, and it is cheaper than the B47.*

12 Whole numbers and decimals

Practise the maths

People working in the financial sector have to work with both positive and negative numbers. If your bank account has a negative balance, it means that you have spent more money than you had available.

12.1 Working with whole numbers

Worked example:

a Round two hundred and fifty-three thousand, four hundred to the nearest thousand.
b Round your answer to part a to one significant figure.

Answer

a *253 000* *(1 mark)*
b *300 000* *(1 mark)*

To round to **one significant figure** (1 s.f.), look at the digit to the right of the first significant figure (2**5**3 000). If it is 5 or more, round up.

1 Round each of these numbers to the nearest thousand.
a 3200 b 4950 c 15 870 d 124 500 e 9875

2 The table shows the populations of four islands.
Arrange the islands in population order, from highest to lowest.

Island	Population
Grenada	90 998
Tonga	121 986
Jersey	91 743
Isle of Man	76 754

3 Round each of these numbers to one significant figure.
a 32 b 180 c 5750
d 14 500 e 4.6 f 3.56

4 A food manufacturer is asked to supply a quarter of a million burgers.
They supplied 40 500 in the first batch. How many more burgers do they still need to supply?

12.2 Working with negative numbers

Worked example:

The temperature starts at 4 °C. Then it falls by 7 °C.
What is the new temperature?

Answer

−3 °C *(1 mark)*

Use a number line to help you. Start at 4 and count back 7 steps.

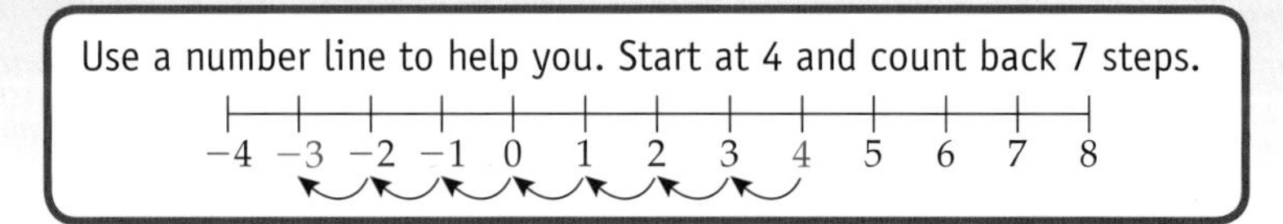

1 What temperature is nine degrees warmer than −5 °C?

2 The table shows the temperatures in four cities.
a Which is the coldest city?
b What is the temperature difference between the coldest and warmest cities?

City	Temperature (°C)
Paris	4
Moscow	−15
New York	2
Helsinki	−2

3 Arrange each set of numbers in order, starting with the lowest.

a 4, −3, 0, 5, −2 b −5, 2, −1, 4, −4

4 Harvey's business account had a balance of −£350.
He has received payments of £480 and £340. What is the new account balance?

12.3 Identifying multiples, factors, primes, squares and cubes

Worked example:

a Write down the first five multiples of 6.
b Write down all the factors of 12.
c Which number in this list is a prime number? 4, 6, 7, 8

Answer

a 6, 12, 18, 24, 30 (1 mark)

b 1, 2, 3, 4, 6, 12 (2 marks)

c 7 (1 mark)

Prime numbers have exactly two factors: 1 and the number itself.
1 and 7 are the only numbers that divide exactly into 7.

1 a Write down the first ten multiples of 3.
b Write down the first ten multiples of 8.
c Which number appears in both your lists from parts a and b?

2 Write down all the factors of

a 15 b 28 c 36 d 60

3 Write down all the prime numbers between 10 and 20.

4 Write down **one** number that is **both** a square number and a cube number.

12.4 Estimating answers

Worked example: Use approximation to estimate the answer to each of these.

a 38×22 b $\frac{8.29 + 6.75}{9.05 - 3.89}$

Answer

a $38 \times 22 \approx 40 \times 20 = 800$ (2 marks)

b $\frac{8.29 + 6.75}{9.05 - 3.89} \approx \frac{8 + 7}{9 - 4} = \frac{15}{5} = 3$ (2 marks)

To **estimate**, round each number to one significant figure. Then do the calculation using your rounded numbers.

Estimate → Calculate → Check it mate!

1 Use approximation to estimate each answer.

a 28×34 b 45×18 c $158 \div 36$ d $872 \div 27$

2 Estimate the answer to each of these calculations.

a 38.4×2.2 b $16.5 \div 3.8$ c $\frac{4.86 + 5.2}{7.05 - 4.76}$ d $\frac{12.8 + 5.59}{2.08 \times 3.91}$

Data sheet

The Reading Festival originates from the National Jazz Festival, which was first held in 1961. The original festival changed names and moved between several sites before settling in Reading in 1971. It is now one of the biggest music festivals in the UK, attracting musicians from all around the world.

The event takes place over the Friday, Saturday and Sunday of the August bank holiday weekend.

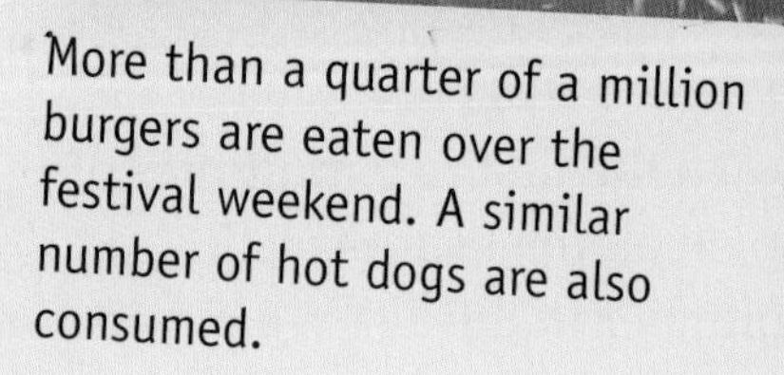

Year	Capacity (number of tickets available)	Ticket price (weekend ticket per person)
1995	45 000	£60
1999	54 000	£78
2000	55 000	£80
2007	75 000	£145
2008	75 000	£155
2009	78 500	£175

2009 prices: tickets	
Weekend ticket (including camping)	£175
Day ticket	£70
Booking fee (per ticket)	£7
Note 1: Booking fee applies to all ticket types	
Note 2: Day ticket holders cannot use the festival campsite even if they have three day tickets	

2009 prices: public transport, parking and camping	
Shuttle bus	£1 single fare
Car park	£5 in advance £10 on gate
Caravan or campervan permit	£20 in advance £30 on gate
Luxury tent (for 1–4 people)	£230
Luxury pod (for 1–3 people)	£235
Sleeping bag	£17.50
Double airbed (inflated)	£20.00
Head torch	£7.00

2009 prices: food and drink

Food Menu	
Burger	£3.50
Pie and mash	£4.50
Sausage and chips	£4.00
Bacon roll	£3.30
Döner kebab and salad	£4.00
Pizza (slice)	£3.50
Energy drink	£2.50

1 How many burgers and hot dogs are eaten over the festival weekend? Give your answer in figures.

2 a What was the ticket holder capacity in 1995?
b Round your answer in part **a** to the nearest ten thousand.
c What is the difference in capacity between 1995 and 2009?

3 A newspaper article in 2009 criticised the Reading Festival organisers and claimed that the price of a weekend ticket had almost trebled in the last 10 years. Was this comment justified? Explain your answer.

4 a How much money would have been generated by ticket sales in 1995, assuming that all tickets were weekend tickets and all were sold?
b How much more money would have been taken in 2009 than in 1995, assuming that all tickets were weekend tickets and all were sold?

5 Luca lives in Reading, within walking distance of the festival site. He has bought a ticket for the Friday night only. During the night he has a burger, a döner kebab and salad, and three energy drinks.
How much has Luca spent in total?

Remember the ticket booking fee!

6 A hot dog is 30p cheaper than a burger. How much money is generated by the sale of hot dogs and burgers over the festival weekend?

Amira, Jay and Tim travel to the festival by campervan. They are hoping to meet up with their friend Oscar who is coming by train.

7 Oscar has a festival ticket for the Sunday. The cost of his return train ticket to Reading station is £85.70.
a Oscar uses the shuttle bus to travel from the station to the festival. How much has he spent in total by the time he arrives at the festival site?
b Oscar has a budget of £178. This must cover the cost that you worked out in part **a**. Which items of food and drink might Oscar buy during the day?

Remember that Oscar must get back to the train station!

8 a Amira, Jay and Tim buy their campervan permit on arrival. How much is the permit?
b Two of the group are staying for the weekend, while Jay only wants to see the acts on Friday night. What is the total cost of the tickets for the group?
c Tim has forgotten to pack his sleeping bag. The group is also told to buy a head torch each if they are staying for the weekend. What is the total cost of these items?
d Jay's friends have suggested that they share all the costs equally, but Jay wonders if he would be better off paying only for what he uses. What would you advise? Explain your answer.

Question key:
Q Open
Beginner
Improver
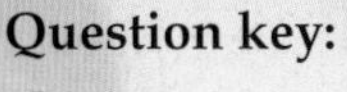
Secure pass

13 Recording data

Practise the maths

The National Grid continuously collects data on the demand for electricity so that it can balance supply with demand on a minute-by-minute basis.

13.1 Recording data in a table

Worked example: Here are the weights (in kg) of a group of people.

62	84	78	45	66	70	79	80	52	54
68	71	75	81	80	70	76	49	63	52

a Put the results into a grouped frequency table using the class intervals $45 \leqslant w < 55$, $55 \leqslant w < 65$, etc.

b How many people's weights are recorded in the table?

c How many of these people weigh less than 65 kg?

Answer

a

Weight (kg)	Tally	Frequency
$45 \leqslant w < 55$	𝍸	5
$55 \leqslant w < 65$	||	2
$65 \leqslant w < 75$	𝍸	5
$75 \leqslant w < 85$	𝍸 |||	8

The heaviest weight is 84 kg so the last interval needed is $75 \leqslant w < 85$.

54 kg is recorded in the $45 \leqslant w < 55$ group.

75 kg is recorded in the $75 \leqslant w < 85$ group.

(3 marks)

b $5 + 2 + 5 + 8 = 20$ people *(2 marks)*

c $2 + 5 = 7$ people *(2 marks)*

1 These are the test scores of a group of students.

15	18	17	10	19	12	20	6	12	9
13	15	19	20	11	15	8	10	14	13
5	2	18	18	14	13	20	20	12	9

a Put the scores into a grouped frequency table using the class intervals 1–5, 6–10, 11–15, etc.

b In which class interval do most scores occur?

c How many test scores are recorded?

d How many students scored 11 or more?

2 Here are the race times (in minutes) of some athletes who competed in the Anglesey half-marathon.

79	74	70	83	90	80	78	83	87	104
94	88	84	81	92	80	100	99	95	86

Design a grouped frequency table to record this data. Choose suitable class intervals.

13.2 Representing data in a pie chart

Worked example: A group of teenagers were asked to choose, from a list, which item of technology they could not do without. The table shows the results.

Draw a pie chart for this data.

Item	Frequency
portable music player	4
mobile phone	15
games console	7
computer	10

Answer

Item	Frequency	Number of degrees
Portable music player	4	$4 \times 10^\circ = 40^\circ$
Mobile phone	15	$15 \times 10^\circ = 150^\circ$
Games console	7	$7 \times 10^\circ = 70^\circ$
Computer	10	$10 \times 10^\circ = 100^\circ$
Total	36	360°

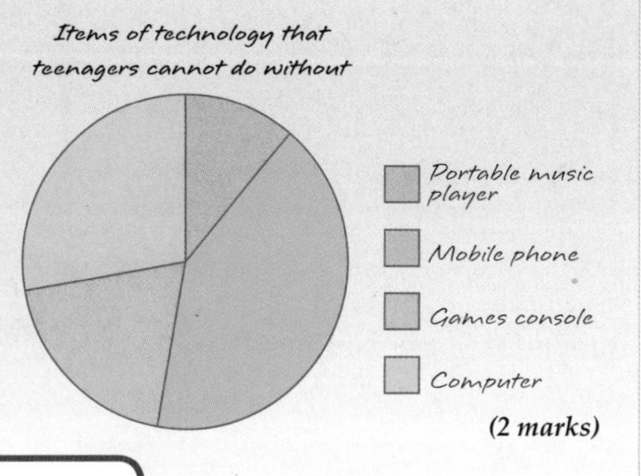

(2 marks)

There are 360° in a full circle. One item is represented by $360^\circ \div 36 = 10^\circ$

1 Lizzie carried out a survey on students' favourite TV programmes.

The results are shown in the table.

Favourite TV programme	Frequency
Dr Who	3
X Factor	9
Hollyoaks	2
Glee	4

Draw a pie chart for this data.

2 The table shows the favourite sporting activities of some people.

Favourite activity	Frequency
swimming	4
rugby	25
orienteering	10
ice skating	6

Draw a pie chart for this data.

13.3 Drawing conclusions from scatter diagrams

Worked example: Here is a scatter diagram. One axis is labelled 'Age of car (years)'.

a What type of correlation does the scatter diagram show? Why?

b Suggest an appropriate label for the vertical axis.

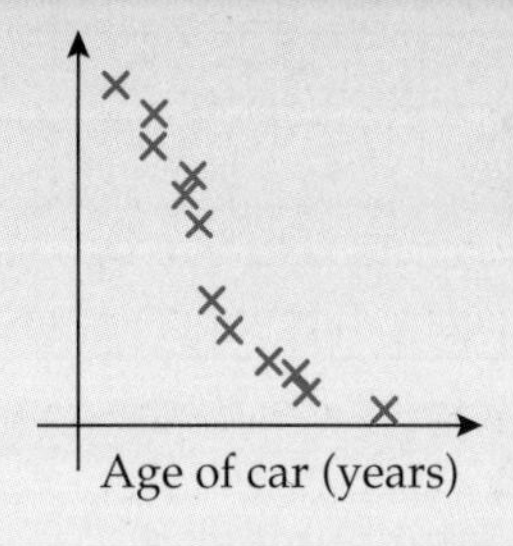

Answer

a *Negative correlation. As one quantity increases, the other decreases.* **(2 marks)**

b *'Value', because a car's value decreases as the car gets older.* **(1 mark)**

There are two other types of **correlation.** Positive correlation indicates that as one quantity increases the other quantity also increases. Zero correlation means that there is no relationship between the two quantities.

1 Which scatter diagram shows the relationship you would expect between

a cigarettes smoked per day and life expectancy

b foot length and shoe size

c sales of ice cream and temperature

d sales of gloves and temperature?

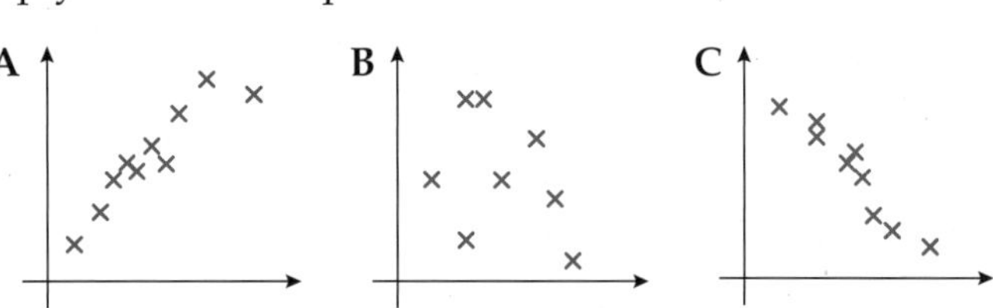

Data sheet

Nathan and Rita are renovating a classic car as part of a college project. Their teacher has helped them set up a blog so that they can produce a record of their progress. Some data on their blog traffic is given below.

In 2008, 346 million people globally were reported to read blogs.

What is a blog?

In simple terms, a blog (or web log) is a type of website on which items are posted regularly. The 'posts' (or entries) are usually arranged in chronological order, with most recent posts at the top of the main page and older entries towards the bottom.

Blogs are usually written by one person and cover a particular topic.

IP address	Return count	IP address	Return count
123.321.123.1	2	544.455.544.5	0
789.987.789.7	1	357.753.357.3	1
345.543.345.3	0	913.319.913.9	4
567.765.567.5	5	791.971.791.7	5
902.209.902.9	4	333.444.555.3	2
119.911.119.1	2	644.466.644.6	1
332.233.332.3	3	369.963.369.3	10
554.455.554.5	7	222.444.666.2	9
987.789.987.9	12	657.756.657.6	0
246.642.246.2	1	853.358.853.8	1

Note: The return count tells you how many return visits have been made from each computer. A return count of '0' means it is a first-time visit.

Day	Total number of visitors	Number of returning visitors
Wednesday	40	14
Thursday	64	24
Friday	82	54
Saturday	110	46
Sunday	86	30
Monday	78	22
Tuesday	50	16
Wednesday	88	48

Visit length (t)	Number of visits
$t < 10$ seconds	120
10 seconds $\leqslant t <$ 30 seconds	72
30 seconds $\leqslant t <$ 5 minutes	110
5 minutes $\leqslant t <$ 20 minutes	46
20 minutes $\leqslant t <$ 1 hour	12

The average number of blog posts globally in a 24-hour period is nine hundred thousand.

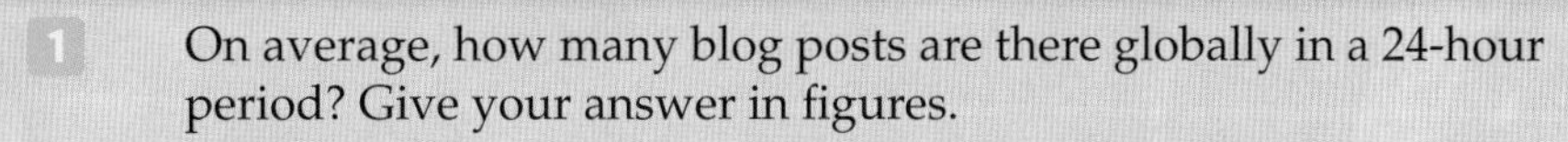

1 On average, how many blog posts are there globally in a 24-hour period? Give your answer in figures.

2 Nathan and Rita post text and photos on to their blog on 21 January 2010, 14 January 2010, 26 January 2010, 28 December 2009 and 29 January 2010.

Arrange these posts in the order that they will be displayed on the blog.

3 How many visitors were there in total to Nathan and Rita's blog over the eight days, Wednesday to Wednesday?

4 Visitors to a blog are classed as either first-time visitors or returning visitors. The total number of visitors is the sum of the numbers of first-time visitors and returning visitors.

Work out the number of first-time visitors to Nathan and Rita's blog over each of the eight days. Record your results in a table.

5 **a** Nathan uses the data they have, together with the results from Q4, to draw a dual bar chart to compare the numbers of first-time visitors and returning visitors. He uses a vertical scale from 0 to 110, going up in steps of 20. Do you think this is a sensible scale? Explain your answer.

b Draw a dual bar chart to show the numbers of first-time visitors and returning visitors to Nathan and Rita's blog over the eight days.

Look at 'Practise the maths' 2.4 for how to draw dual bar charts.

6 Rita sets up a table to record the number and type of visits to their blog, using the data provided for each IP address.

There are 3 computers with a return count of 0, so she records 3 against first-time visits.

There are 13 computers with a return count of 1, 2, 3, 4 or 5, so she records 13 against 1–5 return visits.

Type of visit	Number
first-time visit	3
1–5 return visits	13
5–10 return visits	
10+ return visits	

The IP address is a code given to every computer on the internet. Each IP address is unique.

Nathan says that the table is incorrect. What do you think? Explain your answer.

7 **a** Set up a table with two columns, showing number and type of visit. Record the number of each type of visit to Nathan and Rita's blog using the data provided for each IP address.

b Draw a pie chart to show the information in your table.

c Rita says, 'The pie chart shows that once people visit our blog they keep coming back over and over again.' Is Rita correct? Explain your answer.

8 Look at the data on visit length.

a In which category would a visit of 10 minutes be recorded?

b In which category would a visit of 30 seconds be recorded?

c Draw a pie chart to show the visit length data.

d Can Nathan and Rita use the pie chart as evidence that their blog is a success? Explain your answer.

Question key:

- Q Open
- > Beginner
- >> Improver
- >>> Secure pass

14 Fractions, decimals and percentages

Practise the maths

To arrange the best deal for large purchases, you need to be able to calculate with percentages. Savings interest rates, mortgage rates, loan rates and credit card rates are all given as percentages.

14.1 Working with decimals

Worked example:

a Multiply 12 by 0.4. b Work out £19.20 ÷ 3. c Divide 21 by 0.7.

d Round 6.428 to two decimal places.

Answer

a $12 \times 4 \div 10 = 48 \div 10 = 4.8$ *(1 mark)*

$0.4 = 4 \div 10$ so $12 \times 0.4 = 12 \times 4 \div 10$

b $3\overline{)19.{}^{1}20}$ = £6.40 *(1 mark)*

Line up the decimal point in the answer.

c $7\overline{)210}$ = 30

$21 \div 0.7 = 30$

$21 \div 0.7 = \frac{21}{0.7} = \frac{210}{7} = 210 \div 7$
(multiplying numerator and denominator by 10)

d 6.43 *(1 mark)*

To round to **two decimal places** (2 d.p.), look at the digit in the third decimal place (6.42**8**).

1 Round 2.495 to the nearest whole number.

2 Round each of these numbers to two decimal places.

a 63.535 b 20.082 c 7.1906 d 2.495

3 Work out

a 15.2 + 19.8 b 24.05 + 9.5 c 4 − 3.89 d 48.46 − 29.7

4 A litre of petrol costs 98.5p per litre. What is the cost of 10 litres?

5 Work out

a 24×0.5 b 1.6×0.2 c £68.40 ÷ 4 d $48 \div 0.8$

14.2 Working with fractions, decimals and percentages

Worked example:

a Write 4 as a fraction of 20. b Express 50p as a percentage of £5.

Answer

a $\frac{4}{20} = \frac{2}{10} = \frac{1}{5}$ *(1 mark)*

Always check that your fraction is in its simplest form.

b $\frac{50}{500} \times 100\% = \frac{5}{50} \times 100\% = \frac{1}{10} \times 100\% = 10\%$ *(1 mark)*

Always make sure the quantities are in the same units. £5 = 500p.

1 Arrange these fractions in order of size, from smallest to largest.

$\frac{4}{5}$, $\frac{8}{15}$, $\frac{2}{3}$, $\frac{17}{30}$

First convert the fractions to equivalent fractions, all with the same denominator.

2 Rudi scores eight out of twenty in a test. Work out his score as a percentage.

3 Express 80p as a percentage of £4.

14.3 Calculating percentages

Worked example:

a Work out 17.5% of £80.

b A pair of jeans was originally priced at £70. They are reduced by 20% in a sale. What is the sale price?

Answer

a 10% = £80 ÷ 10 = £8

5% = £8 ÷ 2 = £4

2.5% = £4 ÷ 2 = £2

17.5% = £8 + £4 + £2 = £14 *(2 marks)*

b 80% of £70 = 0.8 × £70 = £56 *(2 marks)*

Always start by finding 10%. You can use this to find other percentages. For example, 5% is half of 10% and 30% is three lots of 10%.

The sale price is (100 − 20)% of the original price, so the sale price is 80% of the original price. You multiply by 0.8.

1 Work out

a 30% of 1200 b 5% of 140 c 15% of 90 d 17.5% of £700

2 The original price of a coat is £60. A reduction of 25% is offered.

a What decimal do you need to multiply by to find the sale price of the coat?

b What is the sale price of the coat?

3 A laptop computer costs £280 (excluding VAT). VAT is 17.5%.
Work out

a the VAT b the total cost of the computer.

VAT stands for Value Added Tax.

14.4 Adding and subtracting fractions

Worked example: Work out

a $\frac{3}{8} + \frac{1}{4}$ b $\frac{2}{3} - \frac{1}{2}$

To add or subtract fractions, the denominators of the fractions must be the same.

Answer

a $\frac{3}{8} + \frac{2}{8} = \frac{5}{8}$ *(2 marks)*

Write $\frac{1}{4}$ as an equivalent fraction with denominator 8. $\frac{1}{4} = \frac{2}{8}$

b $\frac{4}{6} - \frac{3}{6} = \frac{1}{6}$ *(2 marks)*

A common multiple of 3 and 2 is 6. Write both fractions as equivalent fractions with denominator 6. $\frac{2}{3} = \frac{4}{6}$ and $\frac{1}{2} = \frac{3}{6}$

1 Work out

a $\frac{1}{3} + \frac{1}{4}$ b $\frac{2}{5} + \frac{1}{4}$ c $\frac{7}{10} - \frac{3}{5}$ d $\frac{7}{9} - \frac{1}{6}$

2 Nathan eats $\frac{1}{8}$ of a pizza. Emilio eats $\frac{3}{4}$ of the same pizza.
How much of the pizza has been eaten?

3 $\frac{2}{3}$ of a group of people travel to work by train and $\frac{1}{9}$ travel by bus.
What fraction of the group travel by public transport?

Data sheet

When you pay for something using a credit card, the provider (for example, a bank) is lending you the money. Every month you receive a statement which lists all the transactions that you have made with the card. You then have a certain amount of time (for example, 28 days) to make a payment. If you repay the balance in full, no interest is charged.

Jacob has had a credit card for six months. His credit card statement dated 01 February 2010 is shown below.

Ravelo Furniture

31/01/2010

Wooden bookcase	£42.00
VAT @ 17.5%	£
TOTAL	£

In November 2009, outstanding credit card balances stood at £63.5 billion in the UK. This was nearly £3 billion lower than a year earlier.

VAT (Value Added Tax) is a tax that is charged on most goods and services.

***CreditGold* credit card**

Account holder: Mr Jacob Ewing

Statement date: 01 February 2010

Account number: 22** **** **** **89

Minimum payment: £10.00

Payment due date: 01 March 2010

Previous balance: £645.80

Credit limit: £1200

Payment received: £645.80

Date	Transaction details	Amount
06 January 2010	SuperFood Supermarket	£49.42
14 January 2010	Bill's Garage – Diesel	£34.32
17 January 2010	SuperFood Supermarket	£54.70
21 January 2010	Model Clothing	£65.88
22 January 2010	Bill's Garage – Diesel	£28.53
25 January 2010	Games Galore – Online	£48.20
28 January 2010	SuperFood Supermarket	£45.88
29 January 2010	TicketMan – Online	£86.55
31 January 2010	Bill's Garage – Diesel	£37.17
31 January 2010	Ravelo Furniture	£49.35

Around 62% of the adult population in the UK have a credit card.

***CreditGold* default charges**	
Late payment	£12
Over credit limit	£12
Returned payment	£12

Questions

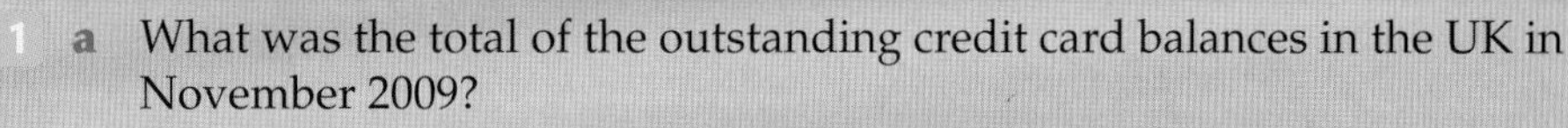

1 **a** What was the total of the outstanding credit card balances in the UK in November 2009?

b Estimate the total amount that was outstanding in November 2008.

2 **a** What percentage of the UK adult population have a credit card?

b Write your answer to part **a** as a fraction in its simplest form.

c Jacob says that 2 in every 5 adults in the UK do not have a credit card. Is Jacob correct? Explain your answer.

Jacob earns £1000 per month after tax working as a sports journalist. He has a CreditGold credit card which has an APR of 18.9%. For the last five months, Jacob has paid off his monthly bill in full.

> APR stands for Annual Percentage Rate. The APR tells you the rate at which you will be charged interest.

3 When his credit card statement arrives, Jacob checks each transaction against the credit card receipt to make sure it is correct. He finds that the receipt from Ravelo Furniture has been torn.

Has Jacob been charged the correct amount for his purchase from Ravelo Furniture? Show your working.

4 Jacob uses a computer spreadsheet to keep track of his credit card spending.

a What total will Jacob enter in his spreadsheet for diesel in January?

b He has entered a total of £134.75 under 'entertainment' for January. Which transactions has Jacob included as 'entertainment'?

c Jacob's budget shows that he can spend up to a quarter of his monthly salary on food. Has he achieved his target in January? Explain your answer.

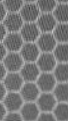

5 **a** What is the total cost of the transactions made by Jacob for the period shown?

b How much will Jacob have to pay so that he does not have to pay any interest?

c If Jacob were to make only the minimum repayment, what would the outstanding balance on his card be?

d Jacob sends a cheque for the total amount owing to CreditGold. The cheque is received on 2 March. The cheque is returned to Jacob as there are insufficient funds in his bank account to pay the total. What charges will be made on his CreditGold card?

> A cheque that is returned because of lack of funds is called a 'bounced' cheque.

6 Assume that Jacob has not yet made any payment to CreditGold. He wants to use his credit card to pay for a holiday booking. The cost of the holiday is £750. Will the payment be authorised by his credit card company? Explain your answer.

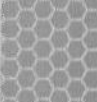

7 What percentage of the total amount owing is the minimum repayment?

> This is the percentage that is used to work out the minimum repayment for this credit card.

8 **a** Did Jacob owe any money on his credit card at the beginning of his February 2010 statement? Explain your answer.

b Can you use your answer to part **a**, together with the previous information you have been given about Jacob, to make any presumptions about his repayment history?

Question key:

Q Open

Beginner

Improver

Secure pass

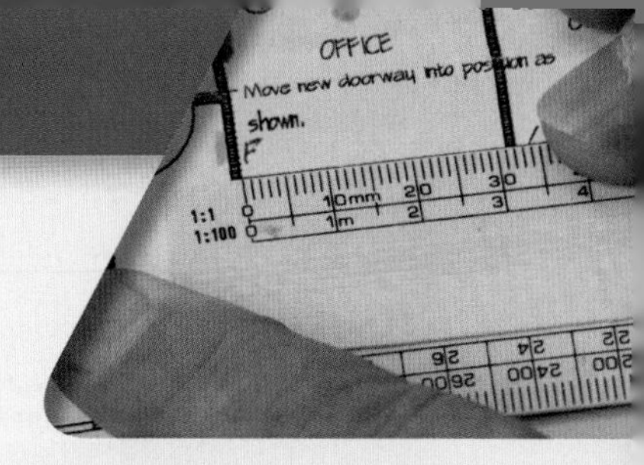

Practise the maths

Ratio is used when business partners need to share their profit fairly. It is also used in scale drawings – from plans of a house to maps of the world.

15.1 Sharing an amount in a ratio

Worked example: Amy has £1.25 and Tom has 75p. They put their money together and buy a lottery scratch card. They win £56.

a Write the amounts spent by Amy and Tom as a ratio in its simplest form.

b Share their winnings in the ratio from part a.

Answer

a *The ratio of the amounts spent is Amy : Tom*

£1.25 : 75p

125 : 75

5 : 3 **(2 marks)**

> To **simplify a ratio** convert both values to the same units, then divide by a common factor.

b *5 + 3 = 8*

£56 ÷ 8 = £7

Amy gets 5 × £7 = £35 Tom gets 3 × £7 = £21 **(3 marks)**

> First work out the total number of parts by adding the numbers in the ratio. Then work out what each part is worth. Finally multiply this value by each of the parts in the given ratio.

1 Write each of these ratios in its simplest form.

a 5 : 20 b 4 : 12 c £6 : £12

d 35p : 55p e 4 cm : 1 m f 20p : £6

> In parts e and f make sure both values are in the same units before simplifying.

2 Share £20 in each of these ratios.

a 4 : 1 b 2 : 3 c 3 : 7

> Set out your workings in the same way as in part b of the worked example.

3 The total cost of a building job is £200.
The ratio of the cost of labour to the cost of the materials is 5 : 3.
What is the labour cost?

15.2 Working out dimensions from scale drawings and maps

Worked example: A scale drawing of the floor plan of a house has a scale of 1 : 40.

a One room in the house is 12 m long. How long is the room on the scale drawing?

b One room on the scale drawing is 20 cm wide. How wide is the room in real life?

Answer

a *12 m = 12 × 100 = 1200 cm*

Scale factor = $\frac{40}{1}$ = 40

1200 cm ÷ 40 = 30 cm

The room is 30 cm long on the scale drawing. **(2 marks)**

> To work out the **scale factor**, divide the larger number in the ratio by the smaller number.

b *20 cm × 40 = 800 cm*

800 cm = 800 ÷ 100 = 8 m

The room is 8 m wide in real life. **(2 marks)**

> To go from real life to the scale drawing, divide by the scale factor. To go from the scale drawing to real life, multiply by the scale factor.

1 A scale drawing of a car has a scale of 1 : 20.

a The real car is 4 m long.
How long is the car on the scale drawing?

b The bonnet of the car is 6 cm long on the scale drawing.
How long is the bonnet in real life?

2 A scale drawing of block of flats has a scale of 1 : 100.

a The block of flats is 54 m high.
How high is the block of flats on the scale drawing?

b The scale drawing shows a fire brigade ladder at maximum vertical reach next to the block of flats.
The ladder reaches 32 cm on the drawing.
How high can the ladder reach in real life?

3 The diagram shows part of a map of a park.
The map has a scale of 1 : 20 000.

a Measure the distance from the West gate to the statue to the nearest mm.

b Work out this distance in real life.

c The park is 1.2 km wide.
How wide is the park on the map?

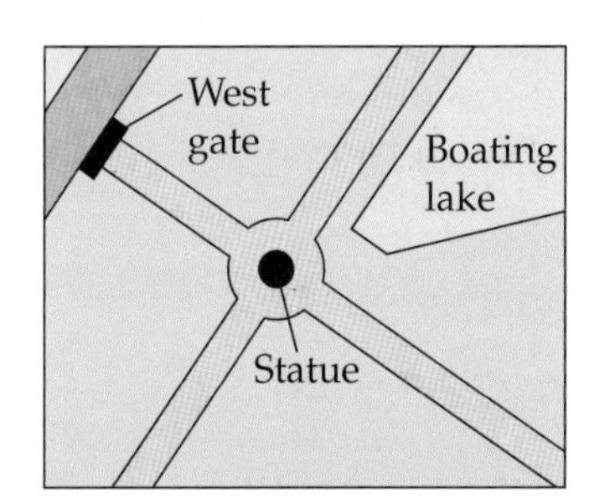

15.3 Using ratio and proportion to solve problems

Worked example: A recipe for risotto uses 300 g of rice for 4 people.
How much rice is needed to make risotto for

a 8 people b 2 people c 6 people?

Answer

a 4 people × 2 = 8 people, so 300 g × 2 = 600 g of rice **(2 marks)**

b 4 people ÷ 2 = 2 people, so 300 g ÷ 2 = 150 g of rice **(2 marks)**

c 2 people × 3 = 6 people, so 150 g × 3 = 450 g of rice **(2 marks)**

1 A recipe for tiramisu uses 32 sponge fingers for 8 people.
How many sponge fingers are needed to make tiramisu for

a 4 people b 16 people c 20 people?

2 A recipe for chocolate cookies uses 150 g of chocolate for 20 cookies.
How much chocolate is needed to make

a 40 cookies b 10 cookies c 30 cookies?

3 Harry is paid £192 for working 24 hours.
How much is Harry paid for working

a 12 hours b 48 hours c 54 hours?

4 Sam's car uses 20 litres of petrol to travel 160 km.
How far will her car travel when it uses

a 5 litres b 25 litres c 32 litres of petrol?

Data sheet

Marco is a catering student. For an assessment he must prepare and cook a three-course meal for 6 people. Here is the menu he chooses.

Menu

Asparagus soup

Noodles with meatballs

Strawberry cheesecake

asparagus soup (serves 2)

75 g almonds
500 ml vegetable stock
5 ml olive oil
2 celery sticks
225 g asparagus

30 ml single cream
15 ml chopped parsley

Preparation time: 15 minutes
Cooking time: 25 minutes

noodles with meatballs (serves 4)

300 g onions
30 ml olive oil
225 g minced beef
30 ml chopped parsley
25 g black olives

30 ml pesto sauce
125 g noodles

Preparation time: 25 minutes
Cooking time: 15 minutes

strawberry cheesecake (serves 12)

100 g butter
150 g plain flour
50 g porridge oats
150 g caster sugar
1.4 kg cottage cheese
4 eggs
2 lemons

120 ml natural yoghurt
450 g strawberries

Preparation time: 30 minutes
Cooking time: 50 minutes
Oven temperature: 180 °C

Top tips

- Soup can be prepared in advance and re-heated before serving.
- One vegetable stock cube will make 200 m*l* of vegetable stock.
- Noodles with meatballs must be served as soon as it is cooked.
- Organic asparagus has a better flavour than ordinary asparagus.
- Pesto sauce can be bought ready-made in a jar.
- Use fresh strawberries on the cheesecake.
- The cheesecake must have time to cool before serving.

1 How much asparagus and single cream does Marco need to make the asparagus soup for 6 people?

2 Marco says, 'I need 8 vegetable stock cubes to make the stock for the soup.'
Is Marco correct? Explain your answer.

3 a How much olive oil does Marco need to make the noodles with meatballs?
b What is the total amount of olive oil that Marco needs for the whole meal?

4 Write a shopping list for Marco that shows the amounts of all the ingredients he must buy to make the cheesecake.

5 Here is an oven temperature conversion table.

Gas mark	°F	°C
1	275	140
2	300	150
3	325	160
4	350	180
5	375	190
6	400	200
7	425	220
8	450	230

Marco is going to cook the cheesecake in a gas oven.
At what gas mark must he cook the cheesecake?

6 Marco must serve the soup at 2 pm, the noodles with meatballs at 2:20 pm and the cheesecake at 2:45 pm.
Write down a plan for Marco, showing when he should prepare and cook the three dishes.

7 Marco has to change his plans as he is asked to make some bread rolls to serve with the soup.
This is the process for making bread rolls.

Step	What to do	Time needed
1	mix and knead the bread dough	10 minutes
2	leave dough to rise	45 minutes–1 hour
3	knead and shape the rolls	5 minutes
4	leave rolls to rise	45 minutes–1 hour
5	bake in oven	15–20 minutes

Bread rolls are best served 10–15 minutes after being taken from the oven.
Write down a new plan for Marco that includes making the bread rolls.

Question key:
Q Open
> Beginner
>> Improver
>>> Secure pass

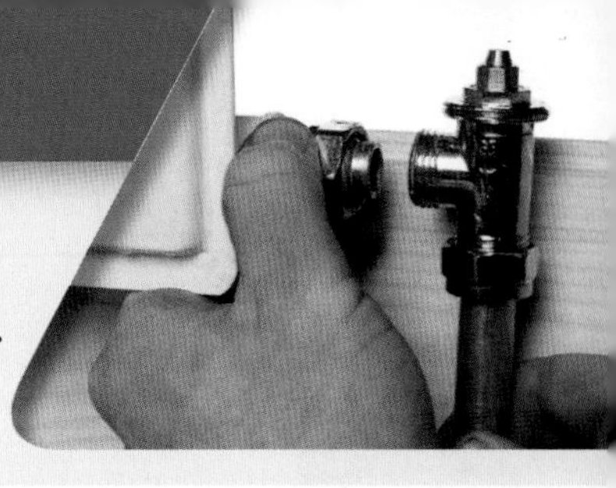

Practise the maths

You need to be able to read tables and graphs in many different jobs. For example, you might need to work out which diameter pipe to connect to a radiator or which size radiator to use in a room.

16.1 Understanding data from stem-and-leaf diagrams

Worked example: The stem-and-leaf diagram shows the test scores of 15 students.

a What was the highest test score?

b Which test score occurred the most often?

c How many students had a score of 25 or more?

Stem	Leaf
0	8
1	2 6
2	1 3 3 3 9
3	0 4 6 7
4	5 8 9

Key
1 | 6 represents 16

Answer

a *49* (1 mark)

b *23* (1 mark)

A score of 23 is recorded three times in the diagram.

c *8 students* (2 marks)

The scores 29, 30, 34, 36, 37, 45, 48 and 49 are all higher than 25.

1 The stem-and-leaf diagram shows the engine capacities of some cars (in litres).

a What is the smallest engine capacity?

b How many cars' engine capacities are recorded in the diagram?

c What is the most common engine capacity?

d How many cars have an engine capacity of less than 2 litres?

Stem	Leaf
0	8 9
1	2 4 5 6 6
2	0 4 5 6
3	0 0 0 2
4	2

Key
1 | 2 represents 1.2 litres

16.2 Using and understanding data from scatter diagrams

Worked example: The scatter diagrams show information about the coffees and hot chocolate sold at Coffee Café.

Sales of latté coffee and price per cup over a two-year period

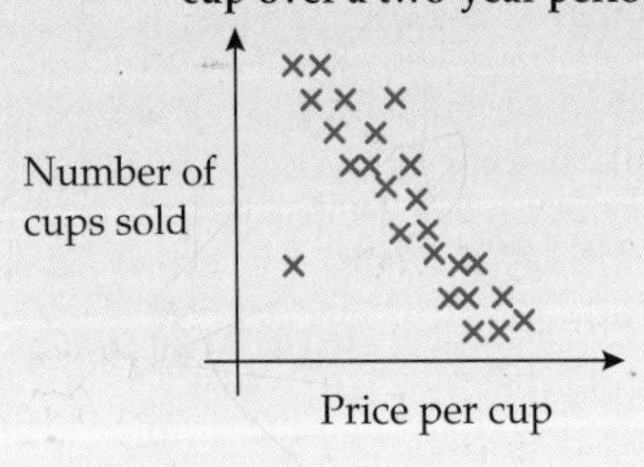

Sales of hot chocolate and price per cup over two winters

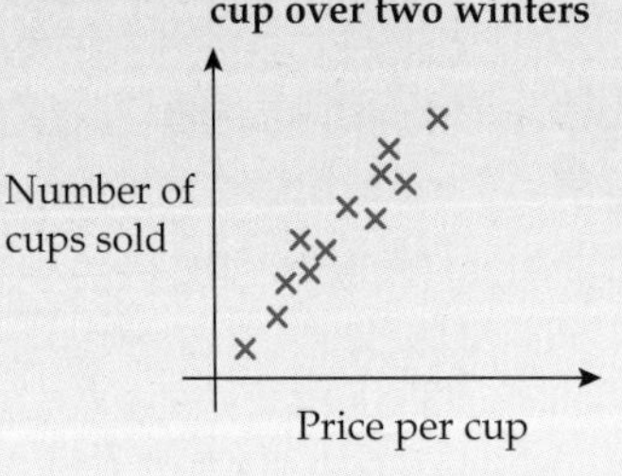

Sales of iced coffee and the number of customers wearing flip-flops over a three-week period

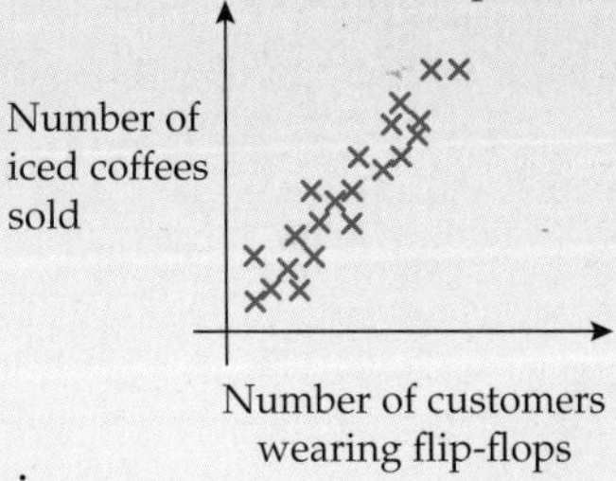

a Use the relevant scatter diagram to describe the general trend in the number of cups of hot chocolate sold.

b What is the statistical term used to describe this trend?

Answer

a *The trend is that the more the hot chocolate costs, the more cups are sold.* (1 mark)

b *Positive correlation* (1 mark)

Interpreting a graph
Always start by reading the title of the graph and then the labels on its axes.

Use the scatter diagrams from the worked example for Q1–3.

1 a Use the relevant scatter diagrams to describe the general trends in the numbers of cups of lattè coffee and of iced coffee sold.

b What is the statistical term used to describe each of these trends?

2 One of the points plotted on the scatter diagram of the number of latté's sold goes against the general trend.

a Explain what this point means.

b Give a possible reason for the point's unexpected position.

3 Do you think that the number of people wearing flip-flops actually caused the increase in the sales of iced coffee, or do you think that something else could have caused an increase in both the sales of iced coffee and the number of people wearing flip-flops? Explain your answer.

16.3 Using and understanding data from dual bar charts and line graphs

Worked example: The dual bar chart shows information about world coffee production in 1991 and 2009.

a In 1991, what percentage of coffee was produced in Other Latin American countries?

b In 2009, is a bag of coffee beans more likely to have come from Africa or Colombia?

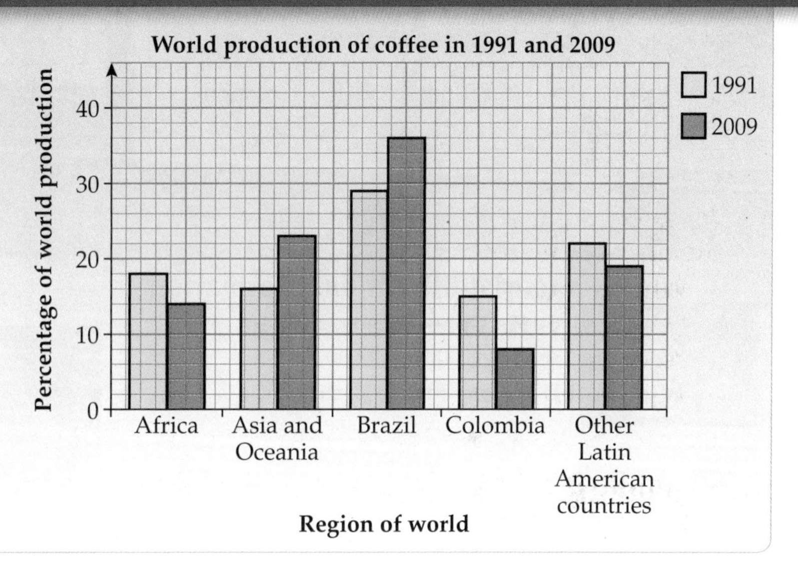

Answer

a *22%* (1 mark)

b *Africa* (1 mark)

1 Use the dual bar chart from the worked example for this question.

a By what percentage share did Brazil's coffee production increase between 1991 and 2009?

b Can you tell from the dual bar chart whether Brazil produced more coffee in 2009 than in 1991? Explain your answer.

2 The dual line graph shows information about the prices paid to coffee growers.

Average prices paid to coffee growers in Colombia and Tanzania, 2000–2008

Price paid (US cents per lb of coffee): 20, 40, 60, 80, 100, 120

Year: 2000, 2002, 2004, 2006, 2008

Colombia, Tanzania

a How much were the coffee growers in Colombia paid per pound for their coffee in 2001?

b What was the difference between the prices paid to coffee growers in Tanzania and Colombia in 2008?

c Between which two years was there the greatest increase in the prices paid to the coffee growers?
Explain how you used the graph to work out your answer.

d Describe how the prices paid to the coffee growers changed over the years 2000 to 2008.

Data sheet

Mrs Kershaw is organising a school trip to Paris for two teachers and 12 sixth-form students.
She plans to arrive in Paris on Monday 10 May in the evening, and to leave on Friday 14 May in the morning.

Paris sightseeing bus

Type of ticket	Adult	Child (under 11)
1 day	€29	€15
2 day	€32	€15

UNLIMITED USE
Hop on/hop off

Seine boat trips – ticket prices

Length of trip	Departs	Adult	Child (under 14)
1 hour	11:00 am	€12	€6
$2\frac{1}{2}$ hours (lunch included)	1:00 pm	€35	€20
1 hour	4:30 pm	€12	€6
3 hours (dinner included)	6:30 pm	€45	€25

Paris bus tour – ticket prices

Length of trip	Departs	Adult	Child (under 11)
$3\frac{1}{2}$ hours	8:30 am	€50	€30
$3\frac{1}{2}$ hours	1:30 pm	€50	€30

Tourist attraction information

Sacre Coeur opening times
Basilica 6:00 am – 11:00 pm *Free*
Dome and crypt 9:00 am – 5:30 pm **€6**

Louvre opening times
9:00 am – 6:00 pm **€9**
6:00 pm – 9:45 pm **€6**
Closed on Tuesdays

Coach trip to the Palace of Versailles
8:45 am – 12:45 pm **€28**
2:00 pm – 6:00 pm **€28**

Eiffel Tower opening times
9:30 am – 11:00 pm **€10**

Parc Astérix theme park opening times
11:00 am – 6:00 pm **€39**

Disneyland Paris theme park opening times
9:30 am – 10:00 pm **€52**

Paris five-day forecast
Monday 10 May – Friday 14 May

	Mon	Tue	Wed	Thu	Fri
Forecast	Sunshine	Cloudy	Cloudy	Rain	Cloudy
Maximum daytime temperature	16°C	14°C	13°C	12°C	15°C

Key
Sunshine
Cloudy

Showers

Rain

1. Which day of the week is forecast to have the lowest maximum daytime temperature?

This is Mrs Kershaw's plan of activities for the students.

Monday:	arrive 6 pm
	evening – visit to Eiffel Tower
Tuesday:	day trip – Parc Astérix
Wednesday:	morning – Paris bus tour
	afternoon – coach trip to the Palace of Versailles
Thursday:	day trip – Disneyland Paris
Friday:	depart 9 am

2. What is the cost per person for a visit to the Eiffel Tower?

3. What is the total cost for the group to go on the Paris bus tour?

4. What is the total cost per person for all the activities that Mrs Kershaw plans?

5. Looking at the weather forecast, do you think that Mrs Kershaw should change her plans?
   Explain your answer.

Three days before they leave, Mrs Kershaw finds out that the Eiffel Tower will be closed on Monday 10 May for maintenance.

She also discovers that the Paris bus tour has been cancelled for the whole week due to safety issues with the buses.

6. Which tourist attraction could the students go to see instead of the Eiffel Tower on Monday evening?

7. Write a new plan of activities for the students for Monday to Friday.
   Your plan must take the following into account:
   - the Eiffel Tower is closed on Monday 10 May
   - the Paris bus tour is cancelled for the whole week
   - the weather forecast
   - the new plan must not cost more than the old plan.

   You must show all your working and explain why you made your decisions.

Question key:

Open   Improver

Beginner   Secure pass

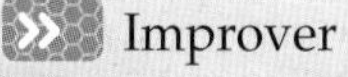

17 Statistical methods

Practise the maths

Statistics are often used in sport – from comparing the average weight of a player in a rugby scrum to looking at how consistently a golfer is playing.

17.1 Calculating the mean, median and mode

Worked example: Ten office workers were asked how much their weekly wage was.
These are their replies.

£330, £200, £170, £200, £350,
£200, £240, £500, £300, £220

a Calculate the mean weekly wage.
b Calculate the median weekly wage.
c Calculate the modal weekly wage.

> The **mean** is the total of the values divided by how many values there are.
>
> The **median** is the middle value when they are put in order of size.
>
> The **mode**, or **modal value**, is the most common value.

Answer

a *Mean wage* $= \dfrac{330 + 200 + 170 + 200 + 350 + 200 + 240 + 500 + 300 + 220}{10}$

$= \dfrac{2710}{10} = £271$ **(2 marks)**

b *In ascending order:*

£170, £200, £200, £200, ***£220, £240****, £300, £330, £350, £500*

The median is half way between £220 and £240. Median = £230 **(2 marks)**

c *Mode = £200* **(1 mark)**

1 Find the mean, median and mode for each of these sets of data.
a 12, 10, 18, 10, 15
b 2.5, 1.0, 0.8, 1.0, 2.5, 1.0, 1.4, 3.8, 2.0, 1.8
c 42, 58, 50, 50, 39, 61, 52, 48

2 Greg went on holiday to Thailand for 10 days. He recorded the temperature at midday each day. These are his results.
33 °C, 32 °C, 34 °C, 36 °C, 35 °C, 32 °C, 33 °C, 32 °C, 32 °C, 31 °C
a Work out the mean temperature.
b Work out the median temperature.
c Work out the modal temperature.

17.2 Deciding which average to use

Worked example: Look at worked example 17.1.
The mean wage is £271, the median is £230 and the mode is £200.
Which average – the mean, median or mode – best describes the data? Explain your answer.

Answer

The median best describes the data.
The mean is usually the best to use, but this data has one extreme value – the £500 is much higher than the rest of the wages. This makes the mean higher than it would normally be, so it is not representative of the data.
There is just one wage lower than the mode but six are higher, so the mode is not representative of the data either. **(2 marks)**

1 a Calculate the mean, median and mode for this set of data.

7, 1, 5, 1, 6, 1, 19, 5

b Which average – the mean, median or mode – best describes the data? Explain your answer.

2 Lynn went to Spain for a week's holiday. She recorded the temperature at 6 pm each day. These are her results.

20 °C, 18 °C, 12 °C, 19 °C, 23 °C, 25 °C, 23 °C

a Calculate the mean, median and mode for this data.

b Which average – the mean, median or mode – best describes the data? Explain your answer.

17.3 Using the mean and range to compare two sets of data

Worked example: Tina and Alyson play 10 rounds of golf.
Tina has a mean score of 78 and a range of 6.
Alyson has a mean score of 76 and a range of 18.
Who is the better golfer? Explain your choice.

Remember that in golf, the lower the score, the better the player.

Answer

I think Alyson is the better golfer because she has the lower mean score.

She has a larger range, which means that she isn't as consistent as Tina, but her mean score is better. **(2 marks)**

The better golfer is up to you – but you **must** give a good reason. You could have given this answer: 'Tina. Her mean is only a little higher than Alyson's, but her range is much lower. So she is the more consistent player.'

1 Isaac can use the number 12 bus or the number 42 bus.
The number 12 has a mean journey time of 28 minutes and a range of 17 minutes.
The number 42 has a mean journey time of 31 minutes and a range of 4 minutes.
If Isaac has to get to his destination in 30 minutes, which bus would you advise him to take? Explain your answer.

2 Cathy and Dianne took part in an experiment on driver reaction times.
The test measured the time it took them to react to hazards in the road and apply their brakes.
Cathy had a mean reaction time of 0.8 seconds and a range of 0.7 seconds.
Dianne had a mean reaction time of 0.93 seconds and a range of 0.3 seconds.
Who do you think is better at reacting to hazards in the road? Explain your answer.

3 A free-range egg farmer needs to buy 2000 hens for egg production.
He wants hens that produce large eggs that are all about the same size.
He finds this information on the internet.
Which type of hen would you advise the farmer to buy?
Explain your answer.

Type of hen	Data on weight of eggs produced	
	Mean	Range
Warren	55 g	7 g
Black Rock	58 g	18 g
Speckled Star	56 g	10 g

Data sheet

Scuba diving is a sport that is becoming increasingly popular in the UK.

John is the diving officer of Llanreath Divers, based in Pembrokeshire. The club has its own boat which it uses to take members out to sea to go diving. Throughout the year, John keeps a record of the number of members who go on each dive.

The members of Llanreath Divers help the environment by collecting the rubbish they find underwater. A lot of the rubbish they find is made of plastic. Plastic bags kill many marine animals every year because the animals eat the bags thinking they are jellyfish.

Plastic nightmare!

Approximately 7800 million plastic bags are used in the UK each year. This is a nightmare for the marine animals that are killed by plastic bags. It also causes problems for local authorities, who are fast running out of landfill space for rubbish.

What is the solution?

The government is considering a tax of 9p per plastic bag, but this raises further questions. Will the money raised from the tax be spent on saving the environment? Will shopkeepers be persuaded to use cornstarch bags instead of plastic bags? If they do, will they charge the customer for the bags they use?

Llanreath Divers information

Number of members	47
Yearly membership fee	£50
Money out – boat expenses	£25 per dive
Money in – charges for members	£12 per dive

Llanreath Divers – dive records 2008 and 2009

	Number of members per dive																		Total number of dives
2008	7	4	8	10	7	6	3	3	7	6	7	4							12
2009	3	5	3	4	7	3	5	4	3	7	5	3	4	5	7	4	6	3	18

1 **a** What was the greatest number of members per dive in 2008?

b What was the range for the number of members per dive in 2008?

c John says, 'The range for the number of members per dive in 2008 was greater than the range for 2009. This means that the number of members per dive in 2008 was more varied than in 2009.'

Is John correct? Explain your answer.

2 **a** Work out the modal number of members per dive in 2008 and in 2009.

b Compare your answers in part **a**, and write a comment to explain what your comparison means.

3 John says, 'The mean number of members per dive in 2008 was 6.'

a What was the mean number of members per dive in 2009?

b Just by looking at the means, is it possible to say in which year more money was paid to the club in diving charges? Explain your answer.

4 **a** Show that the median number of members per dive in 2008 was 6.5.

b Explain what a median of 6.5 means.

5 John works out the profit the club makes by subtracting the cost of using the boat from the total income from yearly fees and diving charges.

In which year did the club make the bigger profit? Show all your workings.

6 John sees a newspaper report on the number of plastic bags used in the UK.

The population of the UK is approximately 60 million.

What is the mean number of plastic bags used per person in the UK?

7 If the government decided to introduce the plastic bag tax, how much money would be raised each year?

8 Shopkeepers can buy 1000 plastic bags for £26.

They can buy 500 cornstarch bags for £75.

A cornstarch bag is not a threat to wildlife as it breaks down in the environment.

a Work out the cost per bag for

i a plastic bag

ii a cornstarch bag.

b Imagine that shopkeepers charge their customers for the bags they use.

If shopkeepers changed from plastic bags to cornstarch bags, how much more per year would an average customer have to pay?

Question key:

Q Open

› Beginner

›› Improver

››› Secure pass

18 Probability

Practise the maths

Teachers and students might use probability to predict what questions are going to be set on next year's exam papers.

18.1 Calculating probabilities

Worked example:

a A fair four-sided dice is rolled.
  i What is the probability of rolling a 1? Give your answer as a fraction.
  ii What is the probability of rolling a 2? Give your answer as a decimal.
  iii What is the probability of rolling a 3? Give your answer as a percentage.

b The fair four-sided dice is rolled twice. What is the probability of getting two 1s?

c All the spade cards are removed from a normal pack.
The remaining cards are shuffled. Julie picks a card at random.
What is the probability that Julie's card is the 2 of hearts?

There are normally 52 cards in a pack, in four equal suits.

Answer

a i $\frac{1}{4}$ *(1 mark)*

There is one 1 out of four possible numbers (1, 2, 3 and 4).

ii $1 \div 4 = 0.25$ *(1 mark)*

iii $\frac{1}{4} \times 100\% = 25\%$ *(1 mark)*

b $P(1) = \frac{1}{4}$

The events are independent, so P(A and B) = P(A) × P(B)

$P(1 \text{ and another } 1) = \frac{1}{4} \times \frac{1}{4} = \frac{1}{16}$ *(2 marks)*

c $\frac{1}{39}$ *(2 marks)*

There are 52 − 13 = 39 cards left in the pack.
Only one of them is the 2 of hearts.

1 Daniel picks a card at random from a shuffled pack of cards.
What is the probability that Daniel's card is
  a a red card? Give your answer as a percentage.
  b a 2? Give your answer as a decimal correct to two decimal places.
  c a diamond? Give your answer as a fraction in its simplest form.

2 All the diamond cards are removed from a normal pack.
The remaining cards are shuffled. Jenny picks a card at random.
What is the probability that Jenny's card is
  a a red card? Give your answer as a fraction.
  b a black 2? Give your answer as a decimal correct to two decimal places.
  c a jack? Give your answer as a percentage to the nearest whole number.

3 Joe is playing a board game using a fair six-sided dice.
To win, he needs to roll a 3 or more on his next throw.
What is the probability that he will win?
Give your answer as a fraction, a decimal and a percentage.

4 A fair six-sided dice is rolled twice.
  a What is the probability of getting two 6s?

The same dice is then rolled three times.
  b What is the probability of getting three 6s?

18.2 Identifying most likely outcomes

Worked example: A bag contains 6 red balls and 4 blue balls.
A spinner has 5 equal sections. 2 of them are red and 3 are blue.
David takes a ball from the bag at random and spins the spinner.
Is he more likely to get a red ball from the bag or a red section on the spinner?

Answer

Probability of red ball $= \frac{6}{10} = \frac{3}{5}$ *(or 0.6 or 60%)*

Probability of red on spinner $= \frac{2}{5}$ *(or 0.4 or 40%)*

$\frac{3}{5} > \frac{2}{5}$, *so David is more likely to get a red ball.*

(3 marks)

Simplify $\frac{6}{10}$ by dividing top and bottom by 2.

To compare the probabilities you need to write them both as fractions with the same denominator, or as decimals or percentages.

1 Bag A contains 3 red balls, 4 blue balls and 5 yellow balls.
Bag B contains 10 red balls, 12 blue balls and 18 yellow balls.
Stuart picks one ball at random from bag A and one ball at random from bag B.
From which bag is Stuart more likely to pick

a a yellow ball b a blue ball c a red ball?

2 These letter cards are shuffled and put into bag P.

M A T H E M A T I C A L

These letter cards are shuffled and put into bag Q.

C A T

From which bag – P or Q – are you more likely to pick a letter A?

3 Which of these events is more likely?

A Getting a 4 or more when you roll a fair six-sided dice.

B Getting a 7 or more when you roll a fair ten-sided dice.

4 The tables give information on the numbers of cups of tea, coffee and hot chocolate sold by two cafés in November.

Tea House	Number of cups
coffee	2880
tea	7200
hot chocolate	4320

Coffee House	Number of cups
coffee	23 544
tea	47 088
hot chocolate	7848

a A customer goes into the Tea House in November.
What is the probability that they buy a cup of tea?

b A customer goes into the Coffee House in November.
What is the probability that they buy a cup of coffee?

c Once a customer has decided where to go, is the customer more likely to buy a hot chocolate from the Tea House or the Coffee House?
Show workings to explain your answer.

Data sheet

Sally is writing a report on the patterns of crime in England and Wales. She reads this newspaper article and then goes online to find out more information.

Crime falls by 5%

It was announced yesterday that police-recorded crimes fell by 5% between 2008 and 2009.

There were

- 6% fewer violent offences against people
- 10% fewer vandalism offences
- 10% fewer vehicle offences
- 1% more burglaries
- 1% more risk of becoming a victim of crime.

Number of recorded robberies in 2009 in the north of England

North-east region	Number of robberies (nearest 10)
Cleveland	400
Durham	170
Northumbria	730
Total	**1300**

North-west region	Number of robberies (nearest 10)
Cheshire	580
Cumbria	50
Greater Manchester	7060
Lancashire	870
Merseyside	1690
Total	**10 250**

Burglary rates in England and Wales, 2002–2009

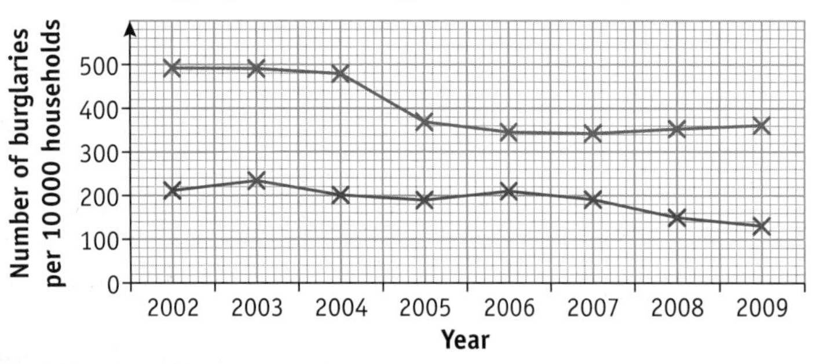

1 **a** By what percentage did police-recorded crimes fall between 2008 and 2009?

b In 2008 there were 2 700 000 vandalism offences. How many fewer vandalism offences were there in 2009?

2 **a** In 2008, what was the urban burglary rate in England and Wales?

b Are you more likely to be burgled if you live in a town (urban) or in the countryside (rural)? Explain your answer.

c Describe how the urban and rural burglary rates in England and Wales have changed over the years from 2002 to 2009.

d In 2009 there were approximately 23 million households in England and Wales.
Work out an estimate for the total number of these households that were burgled.
You must show all your working.

e Use your answer to part **d** to work out the probability that a household chosen at random was burgled in 2009.

3 In the north of England there are two police force areas, the north-east region and the north-west region.
What percentage of the robberies in the north-west region of England were carried out in Greater Manchester? Give your answer to the nearest 1%.

4 Sally is going to include in her report the details of two robberies in the north-east of England and two in the north-west of England.
She chooses the robberies at random from both areas.

> Give all your answers in Q4 correct to two decimal places.

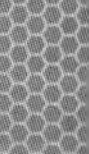

a What is the probability that the first robbery she chooses in the north-east of England is in Cleveland?

b What is the probability that the first robbery she chooses in the north-east of England is in Cleveland and the first robbery she chooses in the north-west of England is in Merseyside?

c What is the probability that the first robbery she chooses in the north-east of England is in Cleveland and the second one is in Durham?

Question key:

Q Open

Beginner

Improver

Secure pass

19 Measures

Practise the maths

When you set up a market stall, you need to be able to convert between metric and imperial measures in order to give equivalent prices. When you go on holiday, you need to be able to convert from one currency to another.

19.1 Converting between metric and imperial units

Worked example: Use the information in the box to complete these approximate conversions between metric and imperial units.

a 2 kg ≈ ___ lb　　b 4 inches ≈ ___ cm　　c 8 km ≈ ___ miles

1 kg ≈ 2.2 lb (pounds)	1 inch ≈ 2.5 cm	1 litre ≈ 1.75 pints
1 lb ≈ 450 g	1 foot ≈ 30 cm	1 gallon ≈ 4.5 litres
1 oz (ounce) ≈ 25 g	1 mile ≈ 1.6 km	

These are the same conversions as you need to know for GCSE maths.

Answer

a 1 kg ≈ 2.2 lb so 2 kg ≈ 2 × 2.2 = 4.4 lb *(2 marks)*

b 1 inch ≈ 2.5 cm so 4 inches ≈ 4 × 2.5 = 10 cm *(2 marks)*

c 1 mile ≈ 1.6 km so 8 km ≈ 8 ÷ 1.6 = 5 miles *(2 marks)*

First decide which conversion you need to use, then decide whether you need to multiply or divide by the conversion factor.

1 Copy and complete these approximate conversions.

a 4 kg ≈ ___ lb　　b 3 oz ≈ ___ g　　c 8 inches ≈ ___ cm

d 3 feet ≈ ___ cm　　e 2 miles ≈ ___ km　　f 4 gallons ≈ ___ litres

g 5 lb ≈ ___ g　　h 8 litres ≈ ___ pints

2 Copy and complete these approximate conversions.

a 15 cm ≈ ___ inches　　b 10.5 pints ≈ ___ litres　　c 150 cm ≈ ___ feet

d 24 km ≈ ___ miles　　e 10 lb ≈ ___ kg　　f 900 g ≈ ___ lb

g 27 litres ≈ ___ gallons　　h 150 g ≈ ___ oz

3 Bethan is following a recipe. She needs 8 oz of flour.
How many grams of flour does she need?

19.2 Using conversion graphs

Worked example: This is a conversion graph between British pounds (£) and American dollars ($).
Use the graph to complete these conversions.

a £10 = $___　　b $20 = £___

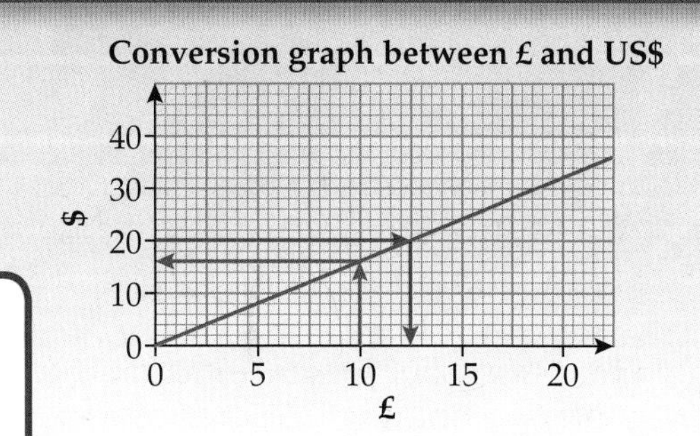

Answer

a £10 = $16 *(2 marks)*

b $20 = £12.50 *(2 marks)*

In part a you need to use the purple line on the graph. In part b you need to use the red line on the graph.

1 This is a conversion graph between British pounds (£) and Mexican pesos (Mex$).
Use the graph to convert

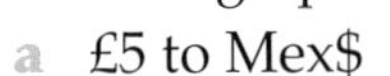

a £5 to Mex$
b £8 to Mex$
c £11.50 to Mex$
d 200 Mex$ to £
e 60 Mex$ to £
f 150 Mex$ to £

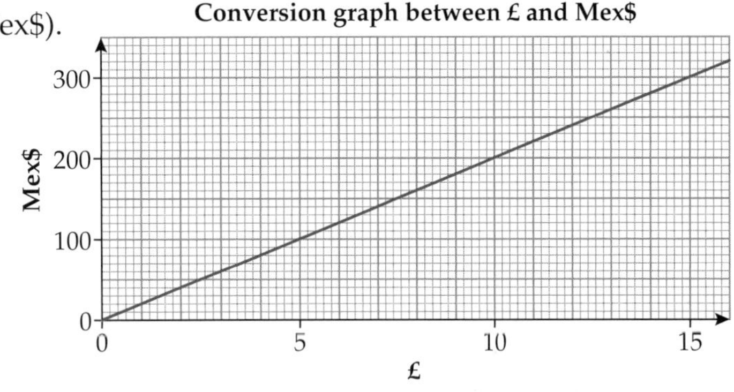

2 When Sian is in Mexico on holiday she buys a hammock which costs 500 Mex$.
How much does the hammock cost in pounds?

19.3 Converting between metric units

Worked example: By converting one measurement to the same units as the other, work out which of each pair is the larger.

a 523 cm or 7.1 m   b 7500 m*l* or 5 litres

10 mm = 1 cm	1000 m*l* = 1 litre	1000 mg = 1 g
100 cm = 1 m	1000 cm^3 = 1 litre	1000 g = 1 kg
1000 mm = 1 m	1000 litres = 1 m^3	1000 kg = 1 tonne
1000 m = 1 km		

Answer

a 523 cm = 523 ÷ 100 = 5.23 m, so 7.1 m is larger than 523 cm. (2 marks)

b 5 litres = 5 × 1000 = 5000 ml, so 7500 ml is larger than 5 litres. (2 marks)

1 By converting one measurement to the same units as the other, work out which of each pair is the larger.
a 721 mm or 59 cm
b 721 mm or 0.75 m
c 75 000 g or 7.2 kg
d 750 kg or 0.7 tonne
e 1550 cm^3 or 2 litres
f 3250 cm or 0.5 km

2 Alison works in a fabric shop.
A customer asks her to cut a length of fabric that is 2.4 m long.
Alison cuts it exactly 2 metres 4 cm long.
Has Alison cut the right length? Explain your answer.

3 How many 500 m*l* bottles can be completely filled from a container holding 2.6 litres of water?

Data sheet

Sharon is a physiotherapist at a gym. Part of her job is to periodically check on the health of the members of the gym.

To do this she looks at a member's

- Body Mass Index (BMI)
- blood pressure.

Feeling the pressure?

Blood pressure is always given as two numbers. It is written like this: 134/88.

The first number is the systolic pressure, which is the pressure when the heart beats.

The second number is the diastolic pressure, which is the pressure when the heart relaxes.

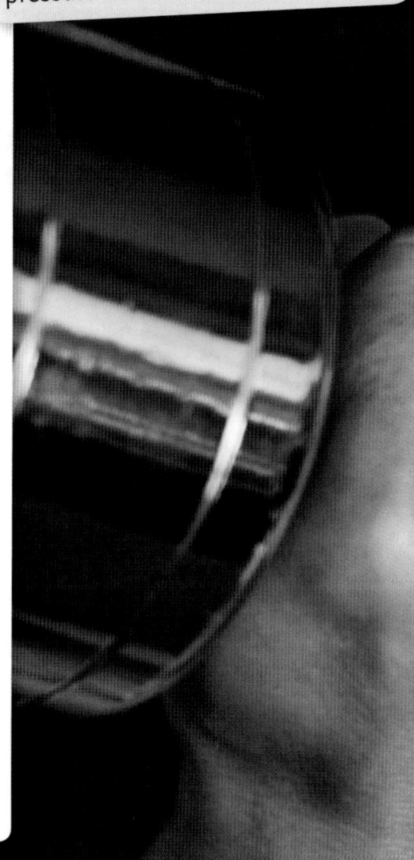

Body Mass Index (BMI) table

Height (m) across; Weight (kg) down

Weight (kg)	1.50	1.54	1.58	1.62	1.66	1.70	1.74	1.78	1.82	1.86
60	27	25	24	23	22	21	20	19	18	17
62	28	26	25	24	22	21	20	20	19	18
64	28	27	26	24	23	22	21	20	19	18
66	29	28	26	25	24	23	22	21	20	19
68	30	29	27	26	25	24	22	21	21	20
70	31	30	28	27	25	24	23	22	21	20
72	32	30	29	27	26	25	24	23	22	21
74	33	31	30	28	27	26	24	23	22	21
76	34	32	30	29	28	26	25	24	23	22
78	35	33	31	30	28	27	26	25	24	23
80	36	34	32	30	29	28	26	25	24	23
82	36	35	33	31	30	28	27	26	25	24
84	37	35	34	32	30	29	28	27	25	24
86	38	36	34	33	31	30	28	27	26	25
88	39	37	35	34	32	30	29	28	27	25
90	40	38	36	34	33	31	30	28	27	26
92	41	39	37	35	33	32	30	29	28	27
94	42	40	38	36	34	33	31	30	28	27
96	43	40	38	37	35	33	32	30	29	28
98	44	41	39	37	36	34	32	31	30	28
100	44	42	40	38	36	35	33	32	30	29

Body Mass Index (BMI) categories

Category	BMI
underweight	<18.5
normal weight	18.5–24.9
overweight	25–29.9
obese: class I	30–34.9
class II	35–39.9
class III	40+

Blood pressure chart for adults aged 30 to 49

Age	Reading	Healthy range
30 to 34	systolic	110 to 134
	diastolic	77 to 85
35 to 39	systolic	111 to 135
	diastolic	78 to 86
40 to 44	systolic	112 to 137
	diastolic	79 to 87
45 to 49	systolic	115 to 139
	diastolic	80 to 88

Questions

1. Sharon works out that Mr Williams has a BMI of 24, which falls in the 'normal weight' category.
   Explain how Sharon arrived at her answer.

2. a What is Mrs Garcia's BMI value?
   b Into which BMI category does Mrs Garcia fall?

3. a Sharon works out that Mr O'Leary needs to lose more than 20 kg to get his weight down into the 'normal weight' category.
   Explain how Sharon arrived at this answer.
   b Sharon notices that Mr O'Leary's systolic and diastolic blood pressure readings are both above the 'healthy range'.
   Explain how Sharon arrived at this answer.

4. Is Mr Williams' blood pressure within the healthy range for his age? Explain your answer.

5. Is Mrs Garcia's blood pressure within the healthy range for her age? Explain your answer.

6. Sharon is writing a report on the health of three members of the gym – Mrs Patel, Ms Evans and Mr McNally.
   She makes recommendations on what each member can do to improve their health.
   What do you think that Sharon may write about each of these three members?
   Explain your answers clearly and show working to support your answers.

Useful conversions

Weight: 1 kg ≈ 2.2 lb
1 stone = 14 lb

Height: 1 inch ≈ 2.54 cm
12 inches = 1 foot

Physiotherapist's records

Name: Sharon Murphy

Date: 06/09/2010

Member's name	Age	Height	Weight	Blood pressure reading
Mr Williams	42	1.82 m	80 kg	128/85
Mrs Garcia	31	1.54 m	64 kg	138/84
Mr O'Leary	48	1.70 m	92 kg	142/92
Mrs Patel	38	1.82 m	60 kg	105/72
Ms Evans	45	1.64 m	74 kg	138/86
Mr McNally	44	6 foot	15 stone	145/90

Question key:
- Q Open
- › Beginner
- ›› Improver
- ››› Secure pass

Practise the maths

You can use letters to represent unknown numbers in formulae. Scientists use a formula for radioactive decay to work out the age of fossilised bones.

20.1 Using letters in formulae

Worked example: A courier company calculates the cost of sending a parcel using this formula. Use this formula to calculate the cost of a parcel weighing 4 kg.

$C = 5W + 8$
C is the cost of sending the parcel in pounds
W is the weight of the parcel in kg

Answer

$C = 5W + 8$ — 5W means $5 \times W$

$= 5 \times 4 + 8$ — Substitute the value for W into the formula. The parcel weighs 4 kg, so $W = 4$.

$= 20 + 8$

$= 28$

The cost is £28. **(2 marks)** — Remember **BIDMAS** (or **BODMAS**): you do **d**ivision and **m**ultiplication before **a**ddition and **s**ubtraction.

1 Work out

a $4 \times 7 + 1$ b $9 - 12 \div 3$ c $10 \times 2 - 6 \times 3$ d $200 \div 10 - 3 \times 4$

2 You can use this formula to calculate the distance you have travelled.

$D = ST$
D is the distance travelled in km
S is your average speed in km/h
T is the time taken in hours

a Calculate the value of D if $S = 30$ and $T = 4$.
b Matt travels at an average speed of 40 km/h for 3.5 hours. How far has he travelled?

3 A phone company uses this formula to calculate the total cost of a bill.

$C = 5T + 15M$
C is the total cost of the bill in pence
T is the number of text messages sent
M is the number of minutes of call time

Calculate the total cost of each of these bills.

a Texts sent: 20
Minutes of call time: 50

b Texts sent: 91
Minutes of call time: 18

c Texts sent: 35
Minutes of call time: 106

20.2 Using formulae with division

Worked example: Jake uses this formula to estimate how much paint he needs for a room. Estimate the amount of paint needed for a room which is 4 m wide.

$V = \frac{W^2}{10}$
V is the amount of paint (in litres)
W is the width of the room (in metres).

Answer

$V = \frac{W^2}{10}$

$= \frac{4^2}{10}$

$= \frac{16}{10}$

$= 1.6$

He needs 1.6 litres of paint. **(2 marks)**

4^2 means '4 squared'. You can use the x^2 button on your calculator.

$\frac{16}{10}$ means $16 \div 10$

1 A swimming pool manufacturer uses this formula to work out how much chlorine to add to a swimming pool.

$C = \frac{N \times L^3}{100}$
C is the amount of chlorine in ml
N is the number of people using the swimming pool
L is the length of the swimming pool in metres

Calculate the amount of chlorine needed if

a 3 people use a 15 m swimming pool

b 5 people use an 18 m swimming pool.

20.3 Using formulae with brackets

Worked example: The total cost of a school theatre trip is calculated using this formula.

$C = N(12 + T)$.
C is the total cost in pounds
N is the number of students
T is the cost of one theatre ticket in pounds

36 students go on the theatre trip.
The cost of one theatre ticket is £8.

Calculate the total cost of the theatre trip.

Answer

$C = N(12 + T)$

$= 36(12 + 8)$

$= 36 \times 20$

$= 720$

The total cost is £720. **(3 marks)**

Remember **BIDMAS** (or **BODMAS**) you work out the **b**rackets first.

1 Alison is working out how much it will cost a group of her friends to go bowling. She uses this formula

$C = N(D + 2G + 5)$.
C is the total cost in pounds
N is the number of friends
D is the cost of dinner per person
G is the cost of one game of bowling per person.

Calculate the value of C if

a $N = 5, D = 12$ and $G = 6$

b $N = 4, D = 15$ and $G = 7.5$

c $N = 7, D = 10.5$ and $G = 8$

Data sheet

Wind turbines create electricity using the power of the wind. Sources of energy like wind and solar power are called renewable sources of energy because they don't use up natural resources like oil or gas.

Around 2% of the electricity generated in the UK comes from wind turbines. This is enough to supply half a million homes.

Power facts

Power output is measured in watts (W). One kilowatt (kW) is 1000 W, and 1 megawatt (MW) is 1 000 000 W.

The power generated by a wind turbine depends on the diameter of the blades and the speed of the wind. You can use this formula to estimate the power of a turbine.

$$P = \frac{D^2 \times W^3}{60}$$

P is the power generated in watts

D is the diameter of the blades in metres

W is the wind speed in mph

Wind farm projects

Project	Bradwell-on-Sea	Whitelee	Camster	Drone Hill
Number of turbines	10	39	25	22
Cost of manufacturing each turbine	£360 000	£620 000	£400 000	£230 000
Cost of transporting each turbine	£80 000	£195 000	£110 000	£65 000
Power output (megawatts)	5	42	15	9

Cost of project $= N(M + T + £10\,000)$

N is the number of wind turbines in the project

M is the cost of manufacturing each wind turbine

T is the cost of transporting each wind turbine

Tempest

Diameter: 3.7 m

Price: **£5875**

Squirrel 150

Diameter: 1.5 m

Price: **£755**

Apollo 1200

Produces 1200 W of power!

Diameter: 2.7 m

Price **£2450**

Questions

1 The UK produces about 45 000 MW of electrical power.

a What percentage of the UK's electrical power comes from wind power?

b Calculate how many MW of electrical power come from wind power.

2 The table opposite shows four planned wind farm projects.

a Use the formula given to calculate the cost of building the Camster wind farm.

b How much more will it cost to build the Whitelee project than the Drone Hill project?

c The Bradwell-on-Sea wind farm project has a budget of £5 million. Will it go over budget? Show all of your working.

3 A newspaper prints a report on wind power.

Daily ★ Record

£1 million per MW

A REPORT ON THE COST OF NEW WIND FARMS

Do you agree with the headline? You must show your working.

4 The world's largest wind turbine is the Enercon E-126. It has a diameter of 126 m. Use the formula given to work out the amount of power produced by the Enercon E-126 if the wind speed is 18 mph. Round your answer to the nearest 100 000 W.

5 Chloe wants to buy a wind turbine for her house. She compares three different products.

a How much more does the Tempest cost than the Squirrel 150?

b Calculate the power output of the Tempest if the wind speed is 12.5 mph. Round your answer to the nerarest watt.

c Calculate the power output of the Squirrel 150 if the wind speed is 15 mph. Round your answer to the nerarest watt.

6 The average wind speed at Chloe's school is 9.5 mph. The manufacturer of the Apollo 1200 claims that it will produce 1200 W of electrical power. Do you agree with this claim? Show all of your working.

Question key:

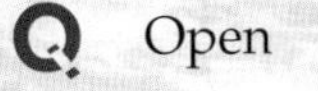

- Q Open
- › Beginner
- ›› Improver
- ››› Secure pass

Data sheet

The Amazon rainforest covers an area of 5.5 million km^2. It is home to millions of different species of plants, animals and insects. However, rainforests are often cut down to provide land for farming and wood for fuel and timber. Around 30 km^2 of Brazilian rainforest are cut down every day.

Brazilian mahogany

- Land in the Brazilian rainforest costs about £160 per hectare.
- One hectare of land might only contain 10 mahogany trees.
- A single tree can have a mass of over 40 tonnes.
- Mahogany wood can sell for £600 per m^3.

Sawn mahogany block

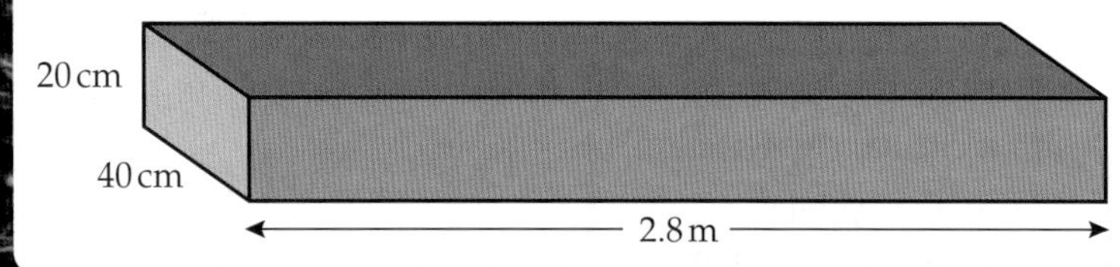

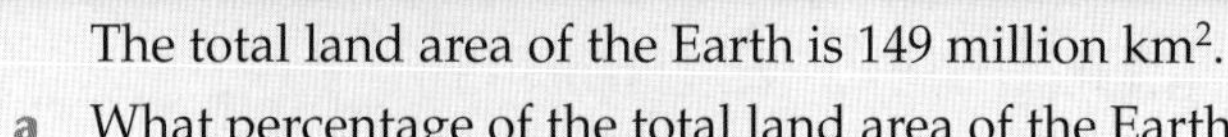

1 The total land area of the Earth is 149 million km^2.

a What percentage of the total land area of the Earth is covered by the Amazon rainforest?
Write your answer correct to two decimal places.

b An environmental charity announces that a million hectares of Brazilian rainforest are being cut down each year. Is this statement accurate?
Show all of your working.

EXAM TIP
There are 100 hectares in 1 km^2.

2 The satellite photo opposite shows a section of rainforest near Porto Velho in Brazil. Two rectangular areas of deforestation have been highlighted.

a What is the perimeter of section A in real life?

b Estimate the total deforested area shown on the photo.
Give your answer in hectares. Show all your working.

3 A beef farming company owns the circle of land shown in red.

a Work out the diameter of this circle of land. Give your answer in km to one decimal place.

b Calculate the circumference of this circle of land.

c Estimate the value of the land owned by the farming company.

d The company wants to build a fence around section B.
Fencing costs £2.20 per metre. They have a budget of £50 000.
Will this be enough? Show all of your working.

EXAM TIP
You can use these formulae:
Area of a circle = $\pi \times$ radius^2
Circumference of a circle = $\pi \times$ diameter

4 Tim says that about a quarter of the land owned by the beef farming company has already been deforested. Do you agree with Tim's statement? Show all of your working.

5 The diagram opposite shows a block of timber cut from a mahogany tree.

a Calculate the volume of this block.

b A single mahogany tree produces 16 blocks of sawn mahogany like the one shown in the diagram. Calculate the total volume of sawn mahogany produced by one tree.

6 The Brazilian government discovers that the farming company has been illegally cutting down mahogany trees on its land. It imposes a fine of £2500 per hectare. Estimate the profit made by the company from 1 hectare of land. Do you think that the fine is a deterrent? Show all of your working.

Question key:
- Q Open
- › Beginner
- ›› Improver
- ››› Secure pass

22 Geometric shapes

Practise the maths

If you are buying or renting a flat you can use a plan to work out where to place your furniture. Architects use computer software which can turn house plans into 3-D models.

22.1 Nets of 3-D objects

Worked example: Sketch a net of this cuboid. Label any lengths on your sketch.

Answer

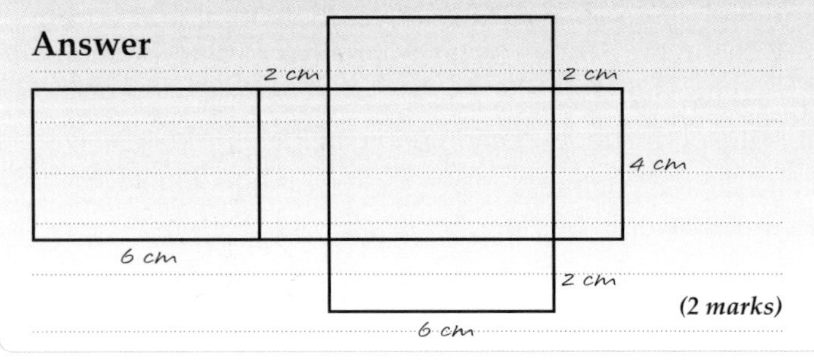

(2 marks)

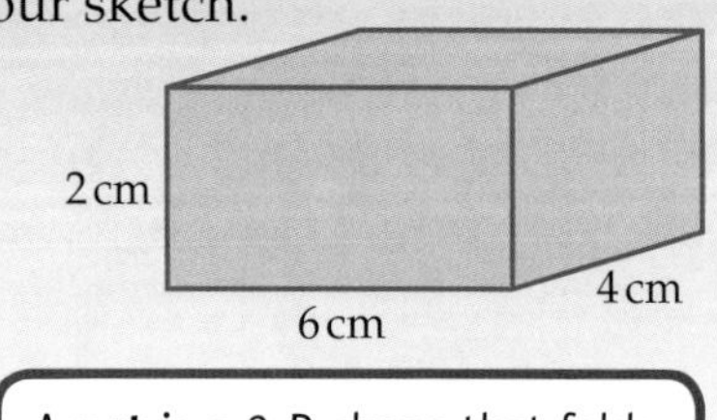

A **net** is a 2-D shape that folds up to make a 3-D object.

1 Which of these shapes could **not** be the net of a cube?

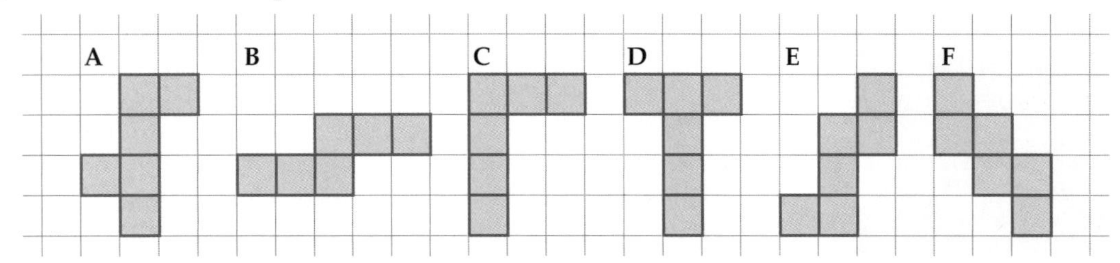

2 Design a net for this cuboid.
Make an accurate drawing of your net.

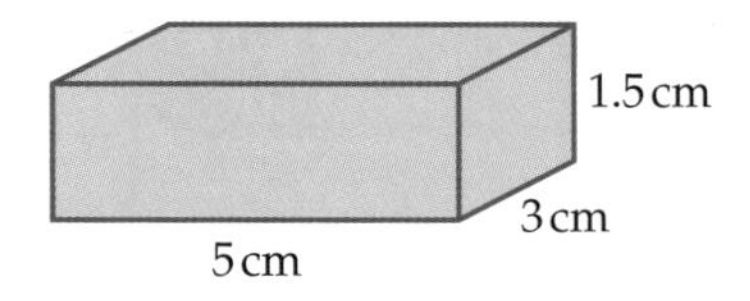

3 The diagram shows a triangular prism.
Draw an accurate net for this shape.

A **triangular prism** has two triangular faces and three rectangular faces. Imagine cutting along the edges of the object and unfolding it.

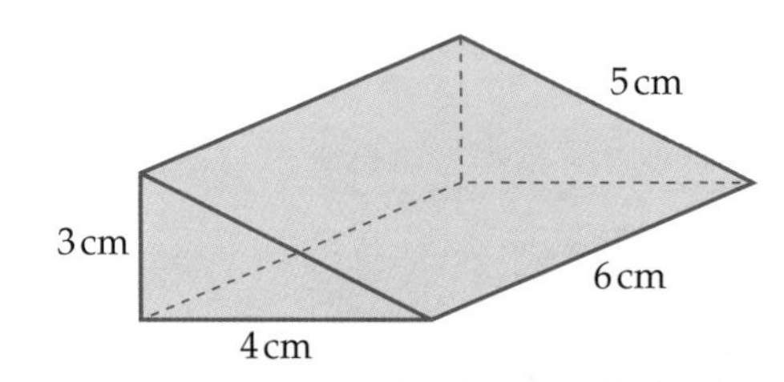

22.2 Drawing 3-D objects

Worked example: The diagram shows a 3-D object drawn on isometric paper.
Draw

a the plan
b the front elevation
c the side elevation.

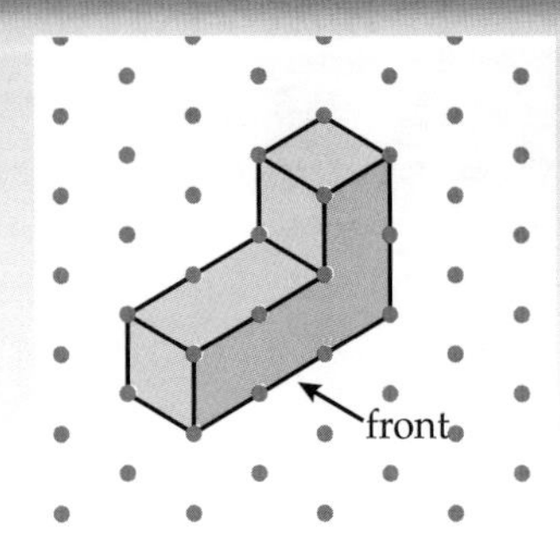

The **plan** is the view from directly above. The views from the front and the side are called the **front elevation** and the **side elevation**.

Answer

a (1 mark)

b (1 mark)

c (1 mark)

1 The diagram shows a 3-D model made from centimetre cubes.
Draw
a the plan
b the front elevation
c the side elevation.

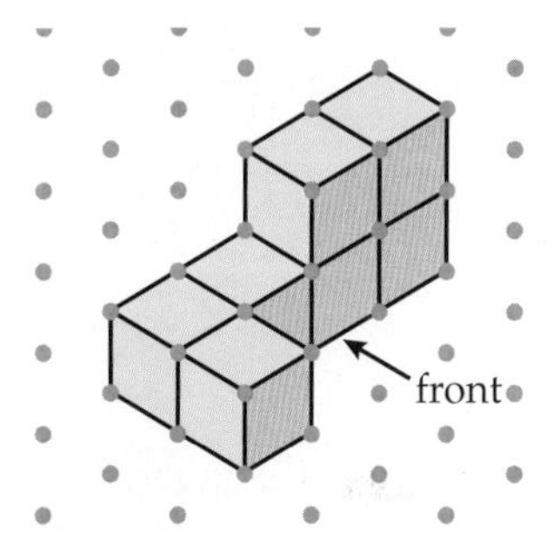

2 Use isometric paper to draw each of these objects.

a

Isometric paper can have lines or dots.

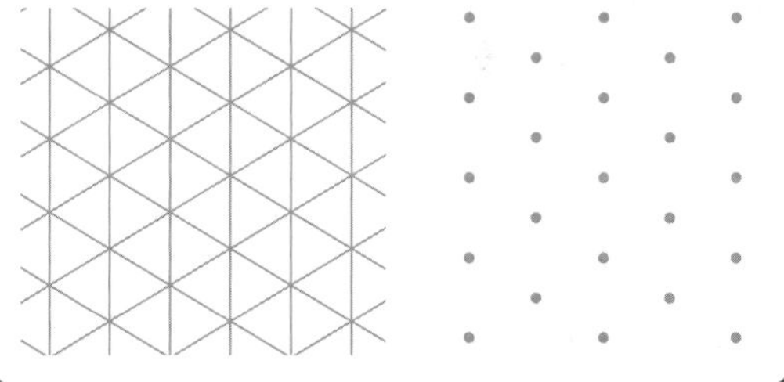

b

c

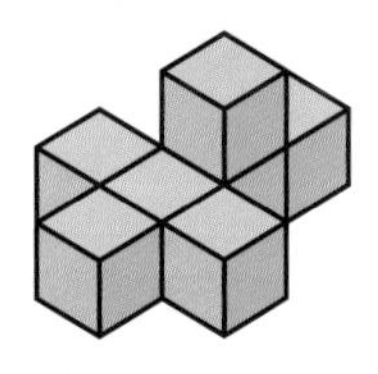

3 The diagram shows the plan and the front and side elevations of a 3-D object.
Draw the object using isometric paper.

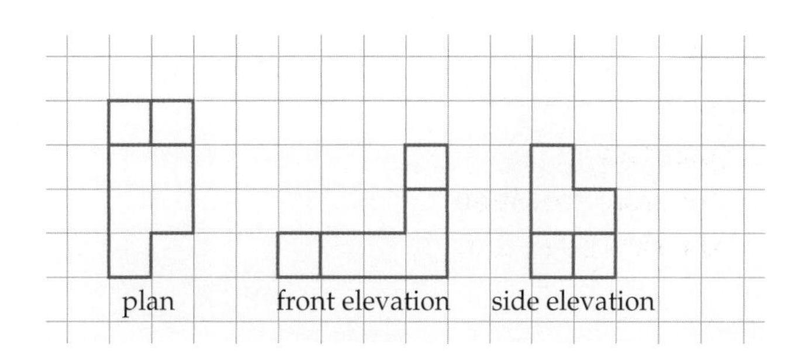

Data sheet

Young Enterprise is a charity which helps students set up and run their own business. Students raise money by selling shares, and then buy or manufacture products to sell.

At Newbridge High School students have decided to start a Young Enterprise business selling candles. They will buy candles from a manufacturer and make decorative boxes to sell them in.

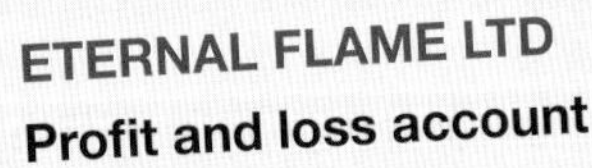

ETERNAL FLAME LTD

Profit and loss account

Money raised from selling shares	£475
Sales	£1220
Total revenue	**£1695**
Cost of candles	£630
Cost of card, glue etc.	£210
Publicity costs	£195
Total expenditure	**£1035**

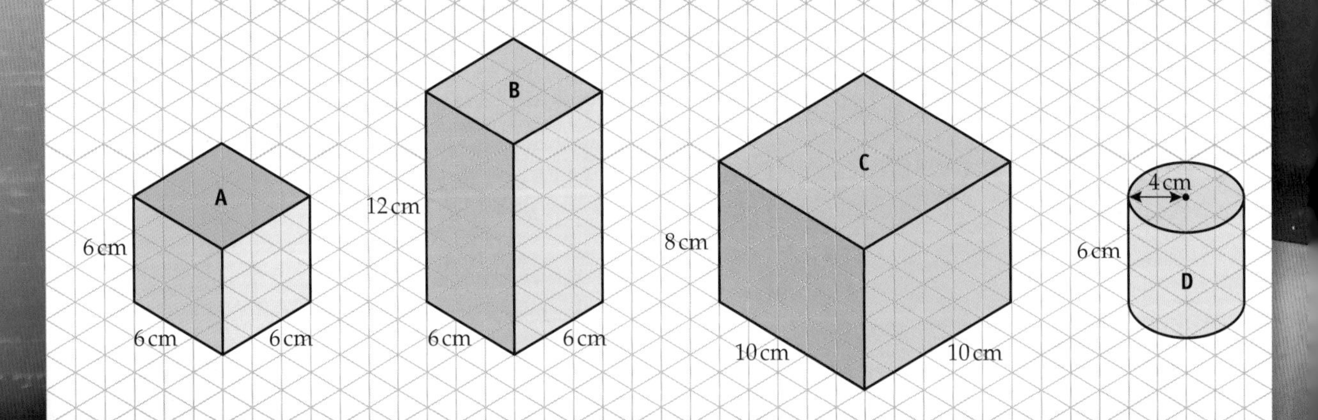

Manufacturer's price list	
A	£0.82
B	£1.03
C	£1.24
D	£0.95
Postage and packing	£9.95

1 Shares in Eternal Flame Ltd cost £5 each.

a How many shares did the company sell in total?

b Calculate the company's profit for the year.

c Young Enterprise companies pay 10% of their profits in tax. How much tax did Eternal Flame Ltd pay in total?

d Helena says that the company can afford to buy back the shares at £7 each. Is she correct? Show all of your working.

2 The company selects four different shapes of candle to sell.

a What is the name of the shape of candle **D**?

b What volume of wax would be needed to make candle **C**?

3 Design a box to hold candle **A**. Draw the net of your box accurately.

4 Amir sketches four possible designs for the net of a box for candle **B**.

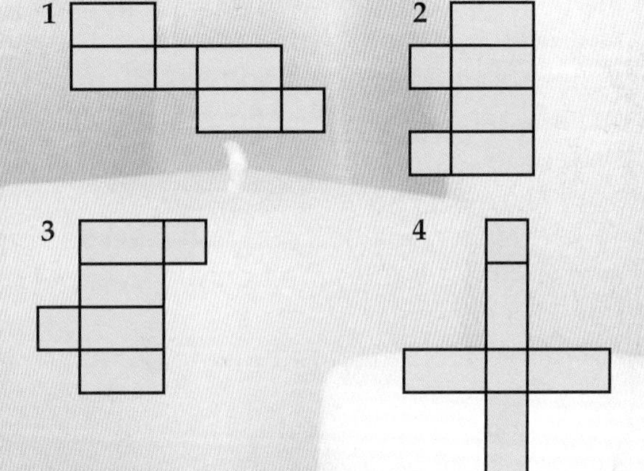

He wants his design to fit on a 30 cm × 30 cm square of card. Which design would you choose? Give reasons for your answer.

5 Jamie orders some candles from the manufacturer. He orders 6 type **C** candles and 15 type **A** candles. The candles arrive in a 30 cm by 40 cm rectangular box.

a How much did this order cost, including postage and packing?

b Show how these candles could be arranged in a 30 cm by 40 cm rectangular box in a single layer. You can use a diagram to explain your answer.

Question key:

Q Open 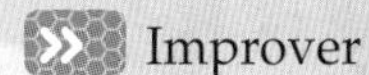 Improver

 Beginner ››› Secure pass

Exam**Café**

Welcome to the Exam Café! This section of the book is here to guide you through the revision period.

First of all remember that examiners are looking to find out what you *can* do, not what you cannot do. The examiner is instructed ALWAYS to mark in a positive manner, so you just need to help him by always showing your working.

Basic exam info

Here's some basic information about the exam.

- For Level 2 you will sit a single Functionality paper in either January, March, June or November.
- The paper lasts 1 hour 30 minutes and has 60 marks.
- There will be between 3 and 5 questions all of which you should attempt.
- Calculators are allowed.
- Each individual question will be based on a single context, e.g. planning a visit to the London Eye, but will be split up into sub-parts.
- Up to 20% of marks are allocated to testing basic maths skills.

Examiner's tips for exam success

1

To be successful in Functional Maths, you have to know your basic maths! You need to be happy with FDPRP:

FRACTIONS, DECIMALS, PERCENTAGES, RATIO AND PROPORTION

Don't forget you will always have your calculator to help. But your calculator is not the answer to everything...

2

Don't depend on your calculator! Even though your calculator will help you to work out your answer, you must follow this rule of thumb:

ESTIMATE – CALCULATE – CHECK IT MATE!

ESTIMATE your answer first, then CALCULATE it using your calculator, then CHECK it using common sense.

WHY SHOULD I SUCCEED AT FUNCTIONAL MATHS

Maths for life
Functional Maths teaches you the maths you will need all the time throughout your life. You can use your maths skills to understand the world around you, or to succeed at work or in education.

MATHS AT HOME
- Choosing the cheapest car insurance – Calculations
- Putting up a wall bracket for a flat screen TV – Angles
- Working out what time to leave home to catch a certain train – Times and timetables

MATHS AT WORK
- Construction engineer – Using ratios
- Architect – Plans and elevations
- Accounts clerk – Adding and subtracting decimals
- Bookmaker – Probability

MATHS IN EDUCATION
- A-level Geography – Reading graphs
- Apprentice electrician – Using formulae
- Biology degree – Negative numbers

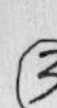

Show your working! Even though you are using a calculator...

YOU MUST SHOW ALL YOUR WORKING OUT

If you make an error, you can still win close to full marks by showing your working and showing the examiner that you understand the process and key steps. Remember: the examiner is always looking to award marks, not take them away.

Use this book! Every chapter starts with...

PRACTISE THE MATHS

This is where you can revise all your basic skills, and make sure you're ready to tackle Functional Maths questions set in context.

Exam**Café**

Data Sheets

All Functional Maths questions are set in everyday contexts. There are 3–5 questions in the exam, and each question comes with a data sheet, providing the material on which you will be tested.

The data sheets for 1 or 2 of the questions will be sent to your school or college 4 weeks before the exam, so you will have some time to prepare those questions in advance!

When you sit the exam you will be given a fresh **data book** that will contain *all* the data sheets for *all* the questions in the exam. This data book will include fresh copies of the data sheets that were sent in advance. You will **not** be allowed to bring your original copies into the exam with you.

Advanced release of data sheets - so what's the best way to prepare?

- A good way to do this is to work with your fellow students and your tutor and try to predict the sort of questions that could be asked from the data sheet.
- Many of the data sheets will have an example, but can you come up with another question that could be asked, and how would you answer it?
- Ask yourself the question: "What is the data telling me?"
- By looking at the information given you can familiarise yourself with the context or ask your tutor to explain it to you if you are unsure.
- Your copy of the data sheet is there to be annotated, but you will not be able to take it into the exam.
- In the exam, you will receive a clean copy as well as further data sheets for the other questions, so not everything can be prepared in advance.

DON'T SPEND TOO MUCH TIME ON THESE SHEETS

– just an hour or two on each one.

Instead, organise yourself a revision schedule so that you

- o revise all the worked examples in this book and the functional skills pages that are dedicated to your level
- o practise exam questions: you will find a Practice Paper at the end of the Exam Café, and your teacher will be able to provide you with more.

What to do before and during your exam

DO	DON'T
Make sure you have pens, pencils, a ruler, a protractor, a compass and a calculator in a see-through bag or pencil case	Don't forget to check that your pens work and your pencils are sharp
Write legibly in a black or blue pen	Don't write in pencil unless you are drawing a diagram or constructing lines or angles
Make sure you have a scientific calculator and know how to use it	Don't forget your calculator or use one you are not familiar with
Think about what you are going to write before you start and keep to the point	Don't "waffle" or write answers if your working doesn't support them
Make sure you attempt each question in order	Don't leave any questions out
Show all your working out in step-by-step, logical calculations	Don't just write an answer without any working or guess answers
Put a line through any working you don't want to be marked	Don't use correction fluid or scribble out answers
Stick to your first answer unless you are sure your original working is wrong	Don't cross out working without replacing it
Make sure your graphs are the right size and you have used a suitable scale on your axes	Don't make your graph too large or too small for the graph paper given
Make notes and underline or highlight the important words in a question	Don't ignore, miss out or misread any of the words in the question
Make sure you have answered every part of the question in the correct order	Don't forget that you may have to write a final sentence to fully answer the question
Read all the information on the data sheet and keep referring back to it	Don't just scan over the data sheet information
Make use of any tables, graphs and charts you are given and annotate them to help you with your calculations	Don't leave a graph or table blank because you are worried about writing something incorrect
Make sure that your answers look sensible	Don't write an answer from your calculator without checking that it makes sense
Go back over your answers if you have any time left at the end of the exam	Don't leave the exam early

Exam Glossary

Functional Maths exams use a particular language. Study the terms below so that you're ready when you sit your exam.

If the question says...	Then you have to...
Analyse	Use the right mathematical methods to carry out step-by-step calculations and check your answer. You might need to use information from the data sheet for this question.
Discuss	Look at the data given to you and show working to support your conclusions. You might need to use information from the data sheet for this question.
Interpret	Use words and diagrams to show how your calculations relate to the question.
Justify your answer	Use words and/or diagrams to support your answer.
Write down	Write down the answer. You don't need to show working for this question, though you might want to anyway!
Give a counter-example	Give an example that shows that the statement in the question is not true. This example could be a value or a diagram.
Make an accurate drawing/use a ruler and compasses	Use a pencil, ruler, protractor and compasses as necessary to construct and measure lines, angles, arcs etc.
Not drawn accurately/ not to scale	Use calculations to find missing lengths and angles. Make sure that you don't measure the lengths and angles on the diagram to find your answer.
Explain	Use words and calculations to explain how you found your answer.
You MUST show your working	Show all of your calculations. If you just give an answer to this question you won't get any marks.
Estimate	Round the numbers given to 1 significant figure. You can use the rounded numbers in your calculations to estimate the answer.
Show	Write down any working or diagrams that are necessary to reach the answer or value given in the question.
Prove	Use logical steps to demonstrate how you have reached your answer. If you use any mathematical facts in your working you must write them down.
Work out	Work out a calculation either mentally, using a written method, or with a calculator.

Calculate	Use a calculator or a written method to find an answer.
Hence	Use the previous answer to help you find a solution.
Hence, or otherwise	Either use the previous answer or show new working to find a solution.
Measure	Use a ruler or protractor to measure a length or angle correct to a given degree of accuracy.
Use the graph	Use the graph given to read off your answer. You can write on the graph to show your working if the graph is in the question. If the graph is only provided on the data sheet, you can still write on it, but remember that the examiner will not see this.
Describe fully	Write down the type of transformation and all the information needed to define it. • Reflection: describe where the mirror line is. • Rotation: write down the centre of rotation, the angle of rotation and its direction. • Translation: write the translation as a column vector. • Enlargement: write down the centre of enlargement and the scale factor of the enlargement.
Comment on	Look at the information in the question and on the data sheet then use calculations and words to write your answer.

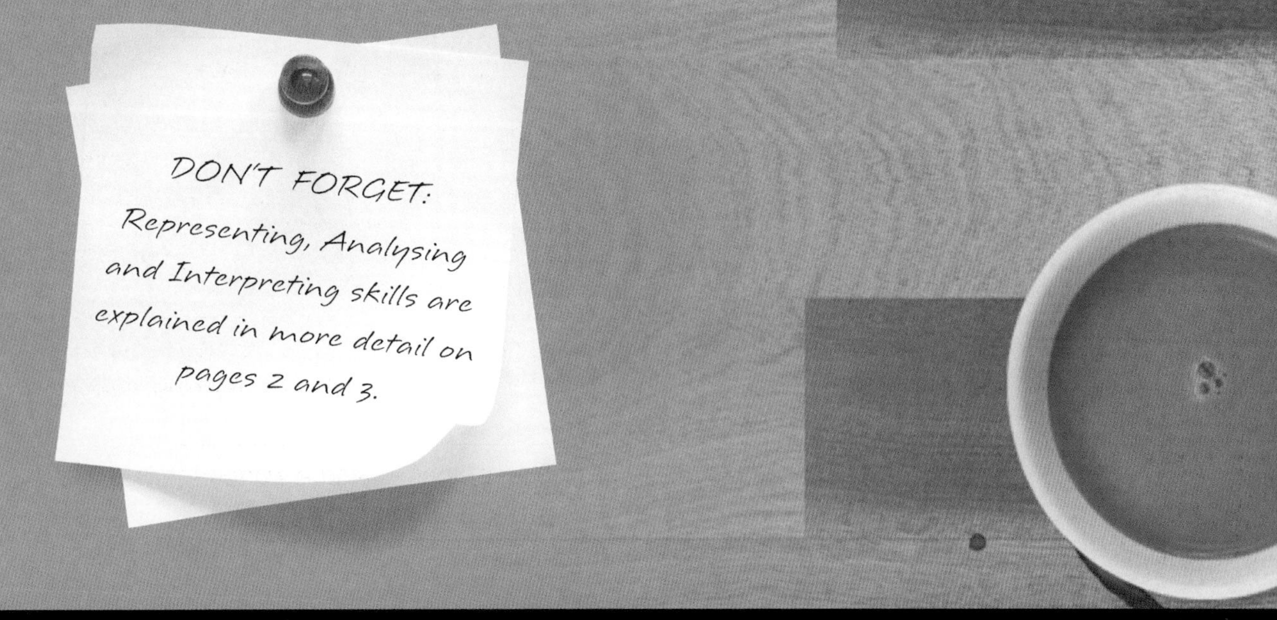

Exam**Café**

Exam question

To make Fizzy Delight, orange juice and lemonade are mixed in the ratio 1 : 3.
Lottie has 200 m*l* of orange juice and 350 m*l* of lemonade.
She wants to use all her orange juice to make Fizzy Delight.
How much more lemonade does she need? **[3 marks]**

Student answers

A

Orange juice = 1, lemonade = 3
350 + 200 = 550

No attempt made to answer the question other than adding the 2 numbers given. No marks awarded.

B

100 ml would require 300 ml therefore 600 ml

One mark awarded for seeing that the amount of lemonade would need to be (3 × 200 =) 600 m*l*.

C

200 ml × 3 = 800 ml
800 − 350 = 450 ml

200 m*l* × 3 gets the first method mark, even though the answer is incorrect. Subtracting 350 gets the second mark, even though the final answer is incorrect.

D

1:3
200:600
600 − 350 = 250 ml
She needs 250 ml more lemonade.

A fully correct answer, showing all working out needed and gaining 3 marks.

Did you find this question difficult?
Have a look at pages 18–21 in Level 1 to revise ratio.
When you're comfortable with that, turn to pages 82–85 in Level 2.

Exam question

Sadiq buys a guitar for £150 and sells it for £210.
Work out his percentage profit. **[3 marks]**

Student answers

A

$210 + 150 = 360$

$\frac{360}{100} = 3.6\%$

The student has added together the two prices. This is an incorrect method and it leads to an incorrect answer. No marks are awarded.

B

$210 - 150 = 60$ so the profit is 60%

The student has attempted the correct calculation (210 – 150) so is awarded 1 mark.

C

$210 - 150 = 70$

$\frac{70}{150} \times 100 = 2.5$

The student has attempted the correct calculation (210 – 150) so is awarded 1 mark even though the answer is incorrect. The student has also attempted to find 70 out of 150 as a percentage. The answer is incorrect, but a second mark is awarded for a correct calculation.

D

$210 - 150 = 60$

$\frac{60}{150} \times 100 = 40\%$

This student has answered the question correctly and is awarded all 3 marks.

AQA Functional Mathematics

Level 2 Practice Paper - 1 hour 30 minutes 60 marks

1 Walking holiday

Data sheet

Steamboat timetable

GR - Glenridding, HT - Howtown, PB - Pooley Bridge, (A) – Arrival, (D) – Departure

	GR (D)	HT	PB (A)	PB (D)	HT	GR (A)
Up to 27 Mar 2010 and 31 Oct 2010–31 Jan 2011	9:45	10:20	–	–	10:20	11:00
	11:10	11:45	12:10	12:10	12:35	1:15
	1:45	2:20	2:45	2:45	3:10	3:50

	GR (D)	HT	PB (A)	PB (D)	HT	GR (A)
29 May–5 Sep 2010	–	–	–	9:45	10:15	10:50
	9:45	10:25	10:50	11:00	11:30	12:05
	10:30	11:10	11:35	11:45	12:15	12:50
	11:15	11:55	12:20	12:30	1:00	1:35
	12:15	12:55	1:20	1:55	2:25	3:00
	1:00	1:40	2:10	2:40	3:10	3:50
	2:05	2:45	3:10	3:20	3:50	4:25
	3:15	3:55	4:20	4:30	5:00	5:35
	4:00	4:35	5:00	5:05	5:30	6:05
	4:45	5:25	5:50	–	–	–

	GR (D)	HT	PB (A)	PB (D)	HT	GR (A)
28 Mar–28 May and 6 Sep–30 Oct 2010	–	–	–	9:45	10:15	10:50
	9:45	10:25	10:50	11:00	11:25	12:00
	11:20	12:00	12:30	12:40	1:10	1:50
	12:15	12:55	1:25	2:00	2:30	3:10
	2:20	3:00	3:30	3:40	4:10	4:50
	3:15	3:55	4:25	4:30	5:00	5:40
	4:50	5:30	6:00	–	–	–

Steamboat ticket prices

Glenridding to Pooley Bridge
(and vice versa)

	Adult	Child	Family
Single	£7.80	£3.90	—
Return	£12.30	£6.15	£29.95

Glenridding or Pooley Bridge to Howtown
(and vice versa)

	Adult	Child	Family
Single	£5.60	£2.80	—
Return	£9.00	£4.50	£24.95

Walkers' Value Ticket – any 3 stages:
Adult £10.70, Child £5.35
Dogs and bicycles £1 per journey

Questions

1 a Lynn, Greg and their dog Polly are on a walking holiday in the Lake District.
The diagram shows the area around one of the lakes, Ullswater.

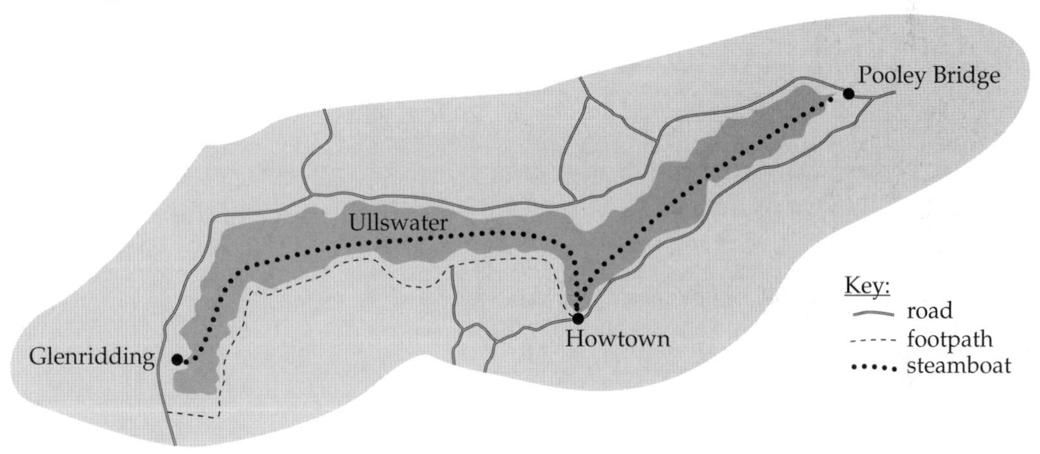

On the 6th of April, Lynn and Greg park their car at Howtown, then walk along the footpath to Glenridding.
They plan to spend about $1\frac{1}{2}$ hours in Glenridding, including having lunch.
Then they plan to catch the steamboat from Glenridding back to Howtown.

Lynn and Greg have a map with a scale of 1 : 25 000.
On their map, the length of the footpath from Howtown to Glenridding is 44 cm.
They walk at an average speed of 4 km/h.

Work out a time plan for Lynn and Greg.
Complete the table showing the time at each stage of their walk.
You must show all your working.

	Time
Leave Howtown	
Arrive Glenridding	
Leave Glenridding	
Arrive Howtown	

(*8 marks*)

b Lynn and Greg have lunch at a café in Glenridding.
This is the menu board.

Lunch menu

Food		Drink	
Baked potatoes with:		Mug of tea	£1.10
Beans and Cheese	£4.20	Mug of coffee	£1.25
Chicken Tikka	£4.70	Hot Chocolate	£1.40
Bolognaise	£4.50	Milkshake	£1.50
		Orange juice	£1.25
Baguette filled with:		Apple juice	£1.25
Sausage or Bacon	£3.25	Lemonade	£1.10
Sausage and Egg	£3.45	Cola	£1.10
Bacon and Egg	£3.45		

Today's special offers:

Buy two baked potatoes and get the cheapest half price.
Buy two filled baguettes and get one mug of tea free.

They order one item of food each and one drink each.
Lynn is a student so they get a 10% discount on the price of the meal.

After lunch they catch the steamboat back to Howtown.
They buy 2 adult tickets and a dog ticket.
They have a maximum of £20 to spend.
Give a suggestion as to what they could order for lunch.
You must show all your calculations.

(7 marks)

2 Travel and climate change

Data sheet

CO_2 emissions per passenger on aeroplane and train journeys

	Return journey by aeroplane		Return journey by train	
Journey from London to:	**Time (hours)**	**CO_2 emissions (kg per passenger)**	**Time (hours)**	**CO_2 emissions (kg per passenger)**
Barcelona	4.5	277	14.25	40
Edinburgh	3.75	194	4.5	24
Nice	4	250	8	36
Paris	3.5	244	2.75	22
Tangier	5	435	30	63

Questions

2 a Aeroplanes and trains emit carbon dioxide (CO_2).
CO_2 emissions are thought to be one of the causes of climate change.
CO_2 emissions from aeroplanes and trains are measured in kilograms per passenger.
The lower the CO_2 emissions per passenger the more environmentally friendly the transport.

Sandra says, 'On average you can cut your CO_2 emissions by about 85% if you travel by train rather than aeroplane'.
Is Sandra correct? You must show workings to justify your answer. (*6 marks*)

b This scatter diagram shows CO_2 emissions per passenger of the train and aeroplane journeys shown in the table on the data sheet.

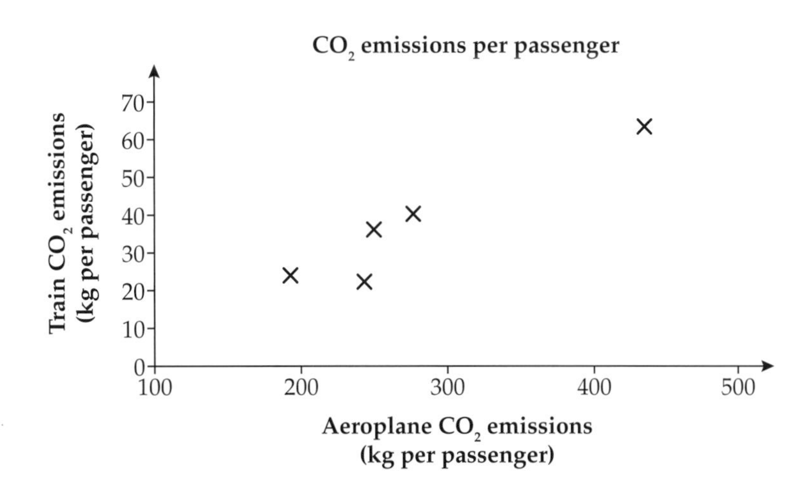

i What type of correlation does this scatter diagram show?
Explain what this correlation means. (*1 mark*)

ii What type of correlation would a scatter graph have that showed the times of the train and aeroplane journeys shown in the table on the data sheet?
You must explain your answer. (*2 marks*)

c This table shows the average number of seats and CO_2 emitted per flight by four common aeroplanes.

Type of aeroplane	Average number of seats used per flight	Average CO_2 emissions per flight (kg)
Airbus A320	120	25 550
Boeing 737	140	29 050
Boeing 747	256	106 800
Boeing 767	212	54 000

The average CO_2 emissions per passenger can be calculated using this formula:

$C = \frac{E}{S}$ where C is the average CO_2 emission per passenger;
E is the average aeroplane CO_2 emission per flight;
S is the average number of seats used per flight.

Sandra works for an international company.
Last year she travelled four times on a Boeing 747.
This year she will be travelling four times on a Boeing 767.
Sandra says, 'I will reduce my CO_2 emissions by just over two-fifths because of the change of aeroplane'.

Is Sandra correct? You must show working to justify your answer. (*5 marks*)

3 Diet and health

Data sheet

The table shows the number of calories per 100 g/ml of different breakfast foods. It also shows a typical portion size.

Food	Calories per 100 g	Typical portion size
bacon	450	50 g
banana	75	100 g
toast	150	150 g
butter	750	20 g
cereal (e.g. cornflakes)	350	30 g
eggs	148	50 g
honey	290	20 g
jam	268	20 g
porridge	370	75 g
sausages	186	200 g
Liquid	**Calories per 100 ml**	**Typical portion size**
black tea or coffee	0	200 ml
apple juice (100 ml)	45	200 ml
orange juice (100 ml)	36	200 ml
milk:		with cereal/with tea or coffee
skimmed (100 ml)	36	300 ml/40 ml
semi-skimmed (100 ml)	50	300 ml/40 ml
full cream (100 ml)	65	300 ml/40 ml

Questions

3 a Gavin is a lorry driver.
He is 28 years old, weighs 15 stone 10 pounds and is 6 feet 2 inches tall.
He wants to lose weight.
To do this he first works out his Basal Metabolic Rate (BMR).

The BMR is the number of calories his body needs each day without taking any exercise into consideration.
He uses this formula:

$$\text{BMR} = 66 + 13.7W + 50.8H - 6.76A$$

where W = weight (kg)
H = height (m)
A = age (years)

Gavin then adds on the calories he uses in an average day.
The table shows what he does in an average day and the number of calories this uses per minute.

Activity	Number of hours	Calories used per minute
Sitting	$4\frac{1}{2}$	1.4
Walking	3	3–6
Lorry driving	6	2.5–4.5
Moderate exercise	$1\frac{1}{2}$	5–7.5

To lose about 1 pound in weight in a week, he must reduce his weekly calorie intake by 3500.

Work out an estimate of the number of calories Gavin should eat per day in order to lose 2 pounds in one week. You must show all your working.

Some useful conversions are:
1 stone = 14 pounds   1 kg $\approx$ 2.2 pounds
1 foot = 12 inches   1 inch $\approx$ 2.5 cm   (*10 marks*)

b Gavin decides to start his diet by eating 3000 calories a day.

He decides to divide his 3000 calories in the ratio 2 : 1 : 3 for breakfast, lunch and dinner.
Plan a breakfast for Gavin that is within the number of calories he is allowed.
You must include a drink and something to eat.
You must show all your calculations.   (*5 marks*)

4 Painting the office

Data sheet

Paint can be bought from the *DIY-help!* shop or the *Mix 'n' Paint* shop.

The prices and the coverage of the paints from each shop are shown here. The paints can only be bought in the quantities shown.

DIY-help!

Matt Emulsion Paint
(1 litre covers 14 square metres)

White
1.0 litre £3.85
2.0 litres £6.75

Green
2.5 litres £21.50
5.0 litres £38.75

Mix 'n' Paint

Matt Emulsion Paint
(1 litre covers 12 square metres)

White
0.5 litre £3.25
1.0 litre £5.25
2.0 litres £7.75

Green
2.5 litres £19.45

Questions

4 a Sham is going to paint his office.
He is going to paint the walls green and the ceiling white.
He is going to give the walls and the ceiling two coats of paint.

Plan of Sham's office

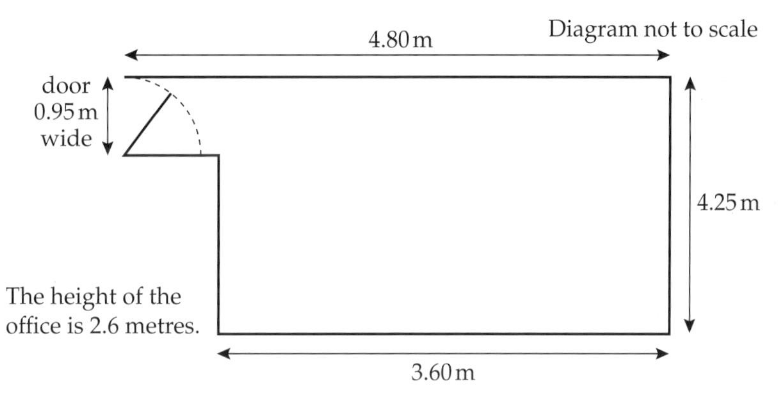

Calculate the amounts of paint that Sham would need from each shop to paint his office.
Show your working clearly and give your answers to one decimal place. (*10 marks*)

b Sham checks his calculations by using a 'paint calculator' on a DIY internet website.

The calculator estimates the amount of green and white paint he needs from each of the shops. The table below gives the results.

	Amount of paint	
	DIY-help!	Mix 'n' Paint
White	2.5	7.5
Green	6.5	3.0

Write down what you notice about the values in the table and your answers to part **a**. Give a possible reason for the differences. (*2 marks*)

c Sham wants to keep his costs as low as possible.
From which shop should Sham buy his paint?
Explain why he should buy the paint from the shop you have chosen and show calculations to justify your answer. (*4 marks*)